AF412925

THE ELEMENTARY DIFFERENTIAL GEOMETRY OF PLANE CURVES

R. H. Fowler, M.A.

Fellow of Trinity College, Cambridge

DOVER PUBLICATIONS, INC.

Mineola, New York

DOVER PHOENIX EDITIONS

Bibliographical Note

This Dover edition, first published in 2005, is an unabridged republication of the work originally published by Cambridge University Press, London, in 1920, as No. 20 of the Cambridge Tracts in Mathematics and Mathematical Physics.

International Standard Book Number: 0-486-44277-2

Manufactured in the United States of America
Dover Publications, Inc., 31 East 2nd Street, Mineola, N.Y. 11501

PREFACE

THIS tract is intended to present a precise account of the elementary differential properties of plane curves. The matter contained is in no sense new, but a suitable connected treatment in the English language has not been available.

As a result, a number of interesting misconceptions are current in English text books. It is sufficient to mention two somewhat striking examples. (*a*) According to the ordinary definition of an envelope, as the locus of the limits of points of intersection of neighbouring curves, a curve is not the envelope of its circles of curvature, for neighbouring circles of curvature do not intersect. (*b*) The definitions of an asymptote—(1) a straight line, the distance from which of a point on the curve tends to zero as the point tends to infinity ; (2) the limit of a tangent to the curve, whose point of contact tends to infinity—are not equivalent. The curve may have an asymptote according to the former definition, and the tangent may exist at every point, but have no limit as its point of contact tends to infinity.

The subjects dealt with, and the general method of treatment, are similar to those of the usual chapters on geometry in any *Cours d'Analyse*, except that in general plane curves alone are considered. At the same time extensions to three dimensions are made in a somewhat arbitrary selection of places, where the extension is immediate, and forms a natural commentary on the two dimensional work, or presents special points of interest (Frenet's formulae). To make such extensions systematically would make the tract too long. The subject matter being wholly classical, no attempt has been made to give full references to sources of information ; the reader however is referred at most stages to the analogous treatment of the subject in the *Cours* or *Traité d'Analyse* of de la Vallée Poussin, Goursat, Jordan or Picard, works to which the author is much indebted.

In general the functions, which define the curves under consideration, are (as usual) assumed to have as many continuous differential coefficients as may be mentioned. In places, however, more particularly at the beginning, this rule is deliberately departed from, and the greatest generality is sought for in the enunciation of any theorem. The determination of the *necessary and sufficient* conditions for the truth of any theorem is then the primary consideration. In the proofs of the elementary theorems, where this procedure is adopted, it is believed that this treatment will be found little more laborious than any rigorous treatment, and that it provides a connecting link between Analysis and more complicated geometrical theorems, in which insistence on the precise *necessary* conditions becomes tedious and out of place, and suitable *sufficient* conditions can always be tacitly assumed. At an earlier stage the more precise formulation of conditions may be regarded as (1) an important grounding for the student of Geometry, and (2) useful practice for the student of Analysis.

The introductory chapter is a collection of somewhat disconnected theorems which are required for reference. The reader can omit it, and to refer to it as it becomes necessary for the understanding of later chapters.

I wish to express my great indebtedness to the Editor, Mr G. H. Hardy, and also to Mr J. E. Littlewood and Dr T. J. I'A. Bromwich, for assistance and advice in the preparation of this tract.

R. H. F.

October 1919.

CONTENTS

CHAPTER I

INTRODUCTION

§ 1·10. We assume in this tract that the reader is acquainted with the ordinary elementary theorems of the differential and integral calculus, as developed, for example, in Hardy's *Pure Mathematics* (2nd Edition, 1914); we apply these theorems to the geometry of plane curves. We shall require more than is there given concerning implicit functions, especially algebraic functions and the curves defined thereby. Such theorems of this type as we require frequently are quoted with references in § 1·50. The more important special properties of algebraic functions are summarized in § 1·60.

We shall use freely the symbols $\sim$, O, o, whose use is now classical, and occasionally $\asymp$ and $\succ$. The reader who is not acquainted with any of them will find the meaning of $\sim$, O, $\asymp$ and $\succ$ explained in Hardy's tract 'Orders of Infinity' (*Cambridge Mathematical Tracts*, No. 12). The definition of o is as follows:

If $f(x)$ and $g(x)$ are any functions of x, and $g(x)$ is positive for all sufficiently large values of x, we write*

$$f(x) = o(g(x)),$$

when $$|f(x)| / g(x) \to 0$$

as $x \to \infty$.

A similar definition applies when x tends to zero, or any other finite limit, instead of to infinity. The introduction of o into Analysis is due to Landau, *vide* Landau, *Handbuch der Lehre von der Verteilung der Primzahlen*, Vol. I, p. 59.

The symmetry of the differential notation and the use of direction cosines are of vital importance in three-dimensional geometry. They can be used with advantage in two dimensions and lend themselves at once to the necessary generalizations. They are therefore used freely here. It is, however, important that the reader should realise the

---

* Alternatively, it will be convenient for our purposes to allow $g(x)$ to be negative instead of positive in the above definition, and also in the definition of O. The only essential requisite in these definitions is that $g(x)$ should not vanish for large values of x.

precise nature of the statement made by a differential formula, and this is frequently emphasised.

A limited selection of examples is given at the ends of the chapters. Besides their more obvious function, these are intended to provide a summary of some of the more important extensions of the theorems proved in the text. References or sketches of a proof are therefore given in such cases, which should enable the reader to complete the proofs.

§ 1·20. **Plane curves.** We regard a plane curve as the locus of points satisfying the equations

$$x = \phi_1(t),$$

$$y = \phi_2(t),$$

for a given range of values of t ($t_0 \leqslant t \leqslant t_1$, say) for which $\phi_1(t)$, $\phi_2(t)$ are continuous single-valued functions of t. A point P on the curve is regarded as identified with a value of t. The variable t is real, and x and y are also always real. We consider throughout only real points and curves.

More information than this about $\phi_1(t)$ and $\phi_2(t)$ will always be required, the amount varying from problem to problem. We may specify conditions to be satisfied by $\phi_1(t)$ and $\phi_2(t)$ either

(1) at a point P, *i.e.* when $t = t_0$,

or   (2) in the neighbourhood of or "near" a point P, *i.e.* in the neighbourhood of t_0,

or   (3) throughout the interval PQ, *i.e.* when $t_0 \leqslant t \leqslant t_1$.

We say that the point $P(t)$ lies between the points $Q_1(t_1)$ and $Q_2(t_2)$ on the curve, when $t_1 < t < t_2$; also that the point $Q(t)$ tends to $P(t_0)$, or $Q \to P$, P and Q being points on the curve, when $t \to t_0$.

A particular case of great importance occurs when $x = t$ or $y = t$, and the curve is given in one of the forms

$$y = \phi_2(x), \ \ x = \phi_1(y).$$

Curves may also be defined by implicit functions. We return to these in § 1·50.

We shall frequently be concerned with straight lines, circles and other curves which depend on certain parameters, and shall study their behaviour as the parameters tend to certain limiting values. In general, suppose the curve is defined by the equations

$$(1 \cdot 21) \qquad x = \phi_1(t, \alpha, \beta, \dots), \ y = \phi_2(t, \alpha, \beta, \dots),$$

or by the equation

$$(1 \cdot 22) \qquad f(x, y, \alpha, \beta, \dots) = 0.$$

Let $a \to a_0$, $\beta \to \beta_0$, ... and suppose that

$$\phi_1(t, a, \beta, \ldots) \to \chi_1(t),$$
$$\phi_2(t, a, \beta, \ldots) \to \chi_2(t),$$
$$f(x, y, a, \beta, \ldots) \to g(x, y).$$

Then we shall say that *the curve $x = \chi_1(t)$, $y = \chi_2(t)$ is the limit of the curve defined by* 1·21 *and that the curve $g(x, y) = 0$ is the limit of the curve defined by* 1·22.

For example we shall define the tangent to any curve at a point P as the limit of the chord PQ as $Q \to P$, the word limit being interpreted in the above sense. The chord may be

$$y = m(\xi)x + c(\xi),$$

where ξ is the parameter of Q. The limit of this chord is

$$y = px + d,$$

where $m(\xi) \to p$, $c(\xi) \to d$ as $\xi \to \xi_0$. Thus, in the case of an algebraic curve such as the above straight line, whose coefficients depend on a parameter or parameters, we regard as the limit of the given curve that curve whose coefficients are the limits of the coefficients of the given curve.

We shall often go further than this and regard the curves defined by 1·21 or 1·22 as approximate representations of their limiting curves. It is then important to be able to describe shortly the closeness or order of such an approximation. Suppose for example that $a = a_0 + \delta a$, and that for any given values of x and y

$$f(x, y, a) = g(x, y) + O(\delta a)^q,$$

as $\delta a \to 0$; in this relation q is a positive integer independent of x and y, while the constant implied by the O may (and in general will) depend essentially on the choice of x and y. Then we shall say that *the curve $f(x, y, a) = 0$, when a is near a_0, represents the limiting curve $g(x, y) = 0$, with an error $O(\delta a)^q$.* When $f(x, y, a)$ is a polynomial in x and y with one non-zero coefficient independent of a, this statement is equivalent to saying that all the coefficients of x and y in $f(x, y, a)$ differ from their limiting values by terms of order $(\delta a)^q$.

§ 1·30. We frequently attempt to impose the minimum conditions that enable us to make a definition or to prove a theorem. In considering the properties of the curve $y = f(x)$ at any point $P(x_0)$ on the curve, we shall always impose the condition that $f'(x_0)$ exists*. This is of course

* See § 2·10, note.

the necessary and sufficient condition for the existence of a tangent not parallel to the axis of y. We then proceed to consider the properties of the curve in the neighbourhood of P, and for this purpose assume in general that $f'(x_0)$ is continuous at P. In this case the curve is always *rectifiable*—it is always possible to assign a meaning to the *length* of an arc of the curve in this neighbourhood. We therefore ignore the question of the necessary and sufficient conditions that a curve should be rectifiable*; a discussion of this question would be out of place here.

Further assumptions are then introduced, such as the existence or continuity of $f''(x_0)$, etc., as required by the problem discussed.

We may mention in passing that the assumption of the existence of $f'(x_0)$ implies rather more than is at once apparent from the definition of a differential coefficient, and that the additional implications are of some geometrical interest (see Note A).

§ 1·40. **Choice of axes. Invariant relations.** We shall assume in general that the axes of coordinates to which our curves are referred are rectangular. It will usually be sufficiently clear to the reader when a theorem remains unaltered by permitting the use of oblique axes.

It is often convenient, in the proof of some general property of a curve, to simplify the proof by referring the curve to a special set of axes, such as the tangent and normal at a point on the curve. It is therefore important to be able to assert that a property proved with a special set of axes is true of curves in general, *i.e.* whatever the axes of reference. *It is permissible to make this assertion owing to the invariance of the formal expressions of lengths and angles for the most general changes of rectangular axes.*

The restriction to rectangular axes is, of course, unnecessary, but a consideration of this case is sufficient for the argument. The general rectangular transformation is

$$x = x' \cos \theta - y' \sin \theta + a,$$

$$y = x' \sin \theta + y' \cos \theta + b,$$

where a, b, θ are any constants. It is easily verified that

$$(x_1 - x_2)^2 + (y_1 - y_2)^2 = (x_1' - x_2')^2 + (y_1' - y_2')^2.$$

This is the property of invariance of length. For the invariance of the (tangent of the) angle between two straight lines, it is easy to see that if the lines are

$$y = m_1 x + c_1, \quad y = m_2 x + c_2,$$

---

* See § 2·50, note.

and they transform into

$$y' = m_1'x' + c_1', \quad y' = m_2'x' + c_2',$$

then
$$\frac{m_1' - m_2'}{1 + m_1'm_2'} = \frac{m_1 - m_2}{1 + m_1 m_2}.$$

Thus all the ordinary metrical properties of curves, which depend on the relations between points on the curves themselves and their tangents and normals, may be established for any system of axes and asserted to hold true in general.

For example, the curvature of a curve is defined in the usual way as

$$\underset{\delta s \to 0}{\mathrm{Lt}} \frac{\delta \psi}{\delta s},$$

where δs is the length of an arc of the curve, $\delta \psi$ the angle between the tangents at the ends of the arc, and δs and $\delta \psi$ are invariant for any change of axes. Hence the above limit, if it exists when the curve is referred to one system of axes, will exist and be equal for all others. It is shown in the usual way that the value of this limit is

$$\frac{d^2 y}{dx^2} \bigg/ \left\{ 1 + \left(\frac{dy}{dx} \right)^2 \right\}^{\frac{3}{2}}.$$

This expression is therefore an invariant for the general change of rectangular axes, as may be directly verified. When we wish to prove properties of the centre of curvature of a point on a curve, we naturally refer the curve to the tangent and normal at the point considered. The value of the above invariant is then $(d^2 y/dx^2)_0$, or $f''(0)$, and the algebra is greatly simplified.

§ 1·50. **Implicit functions.** In addition to the forms of § 1·20, curves may also be defined by implicit functional relations between the coordinates x and y of the type

$$f(x, y) = 0.$$

This case is reduced to the explicit form by the fundamental existence theorem for implicit functions*. We quote here the form most useful for our purposes.

THEOREM 1·51. **Existence theorem for implicit functions.**

Suppose that $F(x, y)$ is a function of the two real variables (x, y) satisfying the conditions:

(1) it is real, one-valued, and continuous, and possesses a continuous partial differential coefficient F_y' in the neighbourhood of (x_0, y_0):

---

* Hardy, *Pure Mathematics*, 2nd Ed., p. 192; Goursat, *Cours d'Analyse Mathématique*, 2nd Ed., Vol. I, Chap. III. The former will in future be referred to as Hardy, *P M.*, and the latter as Goursat, for shortness.

(2) $$F(x_0, y_0) = 0, \quad F_y'(x_0, y_0) \neq 0.$$

Then, (a) there exists a unique function $y = \phi(x)$ which, when substituted in the equation $F(x, y) = 0$, satisfies it identically for all values of x in the neighbourhood of x_0:

(b) $\phi(x)$ is real and continuous in this neighbourhood, and $\phi(x) \to y_0$ as $x \to x_0$.

If further F_x' exists and is continuous in the neighbourhood of (x_0, y_0), the function $y = \phi(x)$ possesses a continuous differential coefficient in this neighbourhood, and

(1·511) $$F_x' + F_y'(dy/dx) = 0,$$

so that

$$dy/dx = \phi'(x) = -F_x'/F_y'.$$

If further all the nth partial differential coefficients of $F(x, y)$ exist and are continuous in the neighbourhood of (x_0, y_0), $d^n y/dx^n$ exists and is continuous in this neighbourhood, and may be calculated by the usual rules.

It may be noted that the extensions of the main existence theorem as quoted assume more than is required about F_x'. In order that $\phi(x)$ may have a differential coefficient at x_0, satisfying equation 1·511, the necessary and sufficient extra condition, over and above the conditions of the main theorem, is simply that F_x' exists at (x_0, y_0). This can be proved by an easy revision of the argument given by Goursat.

More general existence theorems applying to n functions of m independent variables are sometimes required, for example in the theory of contact of curves and surfaces. In such cases the reader is referred to Goursat*.

Cases of exception. If $F_y'(x_0, y_0) = 0$ the theorem breaks down, but if F_x' exists and is continuous, and $F_x'(x_0, y_0) \neq 0$, we can still apply the theorem with the roles of x and y interchanged, and obtain a unique real solution in the form $x = \phi(y)$. It is only if

$$F_y'(x_0, y_0) = F_x'(x_0, y_0) = 0,$$

that the breakdown of the theorem is complete. In this case (x_0, y_0) is called a **singular point**. We shall assume that there are only a finite number of such points in any region with which we deal. The question of the existence of solutions in the neighbourhood of such a point is

---

* See also de la Vallée Poussin, *Cours d'Analyse Infinitésimale*, Vol. I, p. 169. This book will be referred to in future as d.l.V.P. for shortness. References will be given to the 3rd Ed. of Vol. I, and the 2nd Ed. of Vol. II.

discussed by Goursat, p. 102*. We shall return to these points and their geometrical properties in Chapter VI.

A particular choice of axes, referred to which the equation of a curve takes a particularly simple form, is frequently desirable. The validity of the necessary change of axes to convert a general curve into the required particular form may be established as follows.

Let us call for the moment an ordinary point on a curve

$$\text{(i)} \quad y = f(x),$$

a point at which $f'(x)$ exists and is continuous; on a curve

$$\text{(ii)} \quad x = \phi_1(t), \quad y = \phi_2(t),$$

a point at which $\phi_1'(t)$, $\phi_2'(t)$ exist and are continuous and not both zero; and on a curve

$$\text{(iii)} \quad F(x, y) = 0,$$

a point at which F_x', F_y' exist and are continuous and not both zero.

We can then change the origin to the point near which we require the form of the curve, and rotate the axes until in the new system of coordinates (ξ, η)

$$\text{(i)} \; (d\eta/d\xi)_0 = 0, \; \textit{or} \; \text{(ii)} \; (d\eta/dt)_0 = 0, \; \textit{or} \; \text{(iii)} \; (F_\xi')_0 = 0,$$

and also $\qquad\qquad \text{(ii)} \; (d\xi/dt)_0 \neq 0, \; \textit{or} \; \text{(iii)} \; (F_\eta')_0 \neq 0.$

We thus arrive at the following theorem.

THEOREM 1·52. *By the choice of suitable axes a curve, in the neighbourhood of an ordinary point (as defined above), can always be expressed in the form*

$$y = f(x), \; (f(0) = f'(0) = 0).$$

After what precedes this is obvious in case (i) and follows at once from 1·51 in case (iii). In case (ii) 1·51 may be applied to establish the existence of a function ψ such that

$$t = \psi(\xi),$$

where $\psi(0) = t_0$, $\psi'(0) \neq 0$, and $\psi'(\xi)$ is continuous near $\xi = 0$. Theorem 1·52 then follows as before.

The assumptions as to the nature of an ordinary point are more than are necessary for the truth of Theorem 1·52, but we shall only require the theorem in cases in which these assumptions are required for other reasons.

§ 1·60. **Algebraic curves.** If the function $F(x, y)$ is a polynomial in x and y, then y is said to be an algebraic function of x, or the equation $F(x, y) = 0$ is said to define an algebraic curve. This is the most important

* See also d.l.V.P., Vol. II, Chap. IX.

case of an implicit function. The relevant special properties of such curves may be roughly summarized as follows.

(1) Every special assumption that we require to make as to the nature of our curve is always true if the curve is algebraic, with the possible exception of a finite number of points.

(2) In particular there are at most a finite number of points at which

$$F(x, y) = F_x'(x, y) = F_y'(x, y) = 0,$$

near which the form of the curve is not determined by Theorem 1·51. In the neighbourhood of any such point the form of the curve can always be determined (see Chapters VI and VII) by a suitable extension of this theorem.

(3) The study of the form of the curve near infinity, *i.e.* as $x \rightarrow \infty$ or $y \rightarrow \infty$ or both, can always be reduced by a suitable substitution to the study of a similar curve in the neighbourhood of a finite point.

In making the foregoing statements we have tacitly assumed that $F(x, y)$ is not, for instance, of the form $\{G(x, y)\}^2$, where G is a polynomial, for in such a case every point of the curve is a singular point. Such cases of exception are trivial from the present point of view and may be guarded against by restricting ourselves to polynomials which are *irreducible*, that is to say, without factors that are themselves polynomials.

CHAPTER II

THE ELEMENTARY PROPERTIES OF TANGENTS AND NORMALS

§ 2·10. **Definition and existence of the tangent.** DEFINITION. *The **tangent** to a curve at the point P is the limit (if it exists) of the straight line PQ, when $Q \rightarrow P$ along the curve.*

THEOREM 2·11. *The necessary and sufficient condition that the curve $y = f(x)$ should have a tangent at $P(x_0, f(x_0))$, not parallel to the axis of y, is that $f'(x_0)$ should exist. The equation of the tangent is then*

$$y - f(x_0) = f'(x_0)(x - x_0)*.$$

* If we admit infinite differential coefficients, and agree that $f'(x_0)$ has the value $+\infty$ (or $-\infty$) if

$$\underset{x \rightarrow x_0}{\mathrm{Lt}} \frac{f(x) - f(x_0)}{x - x_0} = +\infty \ (\text{or} -\infty),$$

the phrase "not parallel to the axis of y" may be omitted and the words "if $f'(x_0)$ is finite, or $x - x_0 = 0$ if $f'(x_0) = \pm\infty$" inserted at the end of the theorem.

The question is entirely one of phraseology, though perhaps more caution is required if infinite limits are admitted. We shall nowhere admit them in this tract.

(1) The condition is necessary. For if such a tangent exists, the line PQ has a limit of the form $y = Ax + B$. Now, if Q is the point $(\xi, f(\xi))$, the equation of PQ is

$$y - f(x_0) = \frac{f(\xi) - f(x_0)}{\xi - x_0}(x - x_0),$$

and therefore

$$\underset{\xi \to x_0}{\text{Lt}} \frac{f(\xi) - f(x_0)}{\xi - x_0}$$

exists, *i.e.* $f'(x_0)$ exists.

(2) The condition is sufficient. For if $f'(x_0)$ exists, then

$$\frac{f(\xi) - f(x_0)}{\xi - x_0}$$

tends to a finite limit as $\xi \to x_0$, and therefore PQ has a limit, not parallel to the axis of y, which is the tangent at P.

THEOREM 2·12. *In order that the curve*

$$x = \phi_1(t), \ y = \phi_2(t)$$

may have a tangent at the point (t_0), *it is sufficient that* $\phi_1'(t_0)$ *and* $\phi_2'(t_0)$ *should both exist and not both be zero.*

The equation of the tangent is then

$$(2\cdot121) \qquad (y - \phi_2(t_0))\,\phi_1'(t_0) = (x - \phi_1(t_0))\,\phi_2'(t_0)^*.$$

The equation of the chord PQ may be written

$$\frac{\phi_1(t) - \phi_1(t_0)}{t - t_0}(y - \phi_2(t_0)) = \frac{\phi_2(t) - \phi_2(t_0)}{t - t_0}(x - \phi_1(t_0)),$$

which has the required limit as $t \to t_0$, under the stated conditions.

THEOREM 2·13. *In order that the curve*

$$f(x, y) = 0$$

may have a tangent at (x_0, y_0), *it is sufficient that* f_x' *and* f_y' *should be continuous in the neighbourhood of* (x_0, y_0) *and not both be zero at* (x_0, y_0).

The equation of the tangent is then

$$(2\cdot131) \qquad (x - x_0)f_x'(x_0, y_0) + (y - y_0)f_y'(x_0, y_0) = 0.$$

---

* A more useful form in practice is

$$(2\cdot122) \qquad \frac{y - \phi_2(t_0)}{\phi_2'(t_0)} = \frac{x - \phi_1(t_0)}{\phi_1'(t_0)},$$

or more shortly

$$(2\cdot123) \qquad \frac{y - y_0}{y_0'} = \frac{x - x_0}{x_0'}.$$

If one of x_0' or y_0' is zero, the equation must of course be interpreted to mean that the corresponding numerator vanishes.

This is a direct consequence of the Existence Theorem 1·51 and Theorem 2·11. If for example $f_y'(x_0, y_0) \neq 0$, the mere existence of $f_x'(x_0, y_0)$ is sufficient.

§ 2·20. **The tangent as the limit of the chord.** The tangent may be the limit of a chord of more general type than that used in the definition of § 2·10. This is shown by the following theorem.

THEOREM 2·21. *If $f'(x)$ is continuous at x_0, the tangent at x_0 to the curve $y = f(x)$ is the limit of the straight line $Q_1 Q_2$ when Q_1, $Q_2 \to P$ along the curve.*

It is not sufficient here that $f'(x)$ should exist. Something more is required and continuity is a simple and sufficient condition. *If however Q_1, $Q_2 \to P$ from opposite sides*, then the chord $Q_1 Q_2$ tends to the tangent at P provided only that the tangent exists. See Note A.

The equation of the chord $Q_1 Q_2$ is

$$y - f(x_1) = \frac{f(x_2) - f(x_1)}{x_2 - x_1}(x - x_1).$$

The hypothesis of the continuity of $f'(x)$ at x_0 implies the existence of $f'(x)$ at all neighbouring points to P. Hence Q_1 and Q_2 may be taken near enough to P for $f'(x)$ to exist at all points of the interval $x_1 \leqslant x \leqslant x_2$. Therefore by the mean value theorem

$$\frac{f(x_2) - f(x_1)}{x_2 - x_1} = f'(x_1 + \theta(x_2 - x_1)), \quad (0 < \theta < 1).$$

But since $f'(x)$ is continuous at x_0, $f'(x_1 + \theta(x_2 - x_1)) \to f'(x_0)$, when Q_1, $Q_2 \to P$, and the limit of the chord is

$$y - f(x_0) = f'(x_0)(x - x_0),$$

i.e. the tangent at P.

Other forms of the equation of the curve may be treated in a similar manner.

We shall speak of a curve as having *a continuous tangent at P*, when, as $Q \to P$, the tangent at Q tends to the tangent at P. It is easily seen that the necessary and sufficient condition for this in the case of $y = f(x)$ is that $f'(x)$ should be continuous at P, and, for the other two forms, that sufficient conditions are that $\phi_1'(t)$ and $\phi_2'(t)$ should be continuous, or that f_x' and f_y' should be continuous, respectively, at P.

§ 2·30. **Definition and equations of the normal.** DEFINITION. *The* **normal** *to a curve at the point P is a straight line through P at right angles to the tangent.*

The equations of the normal in the three cases studied above are respectively (the axes of coordinates being as always rectangular)

$$(2\text{·}31) \qquad f'(x_0)\,(y - f(x_0)) + (x - x_0) = 0,$$

$$(2\text{·}32) \qquad (y - \phi_2(t_0))\,\phi_2'(t_0) + (x - \phi_1(t_0))\,\phi_1'(t_0) = 0,$$

$$(2\text{·}33) \qquad \frac{x - x_0}{f_x'(x_0,\,y_0)} = \frac{y - y_0}{f_y'(x_0,\,y_0)}\;*.$$

§ 2·40. **The geometrical meaning of differentials.** Consider the curve

$$y = f(x)$$

with a tangent at P, the point (ξ, η), not parallel to the axis of y.

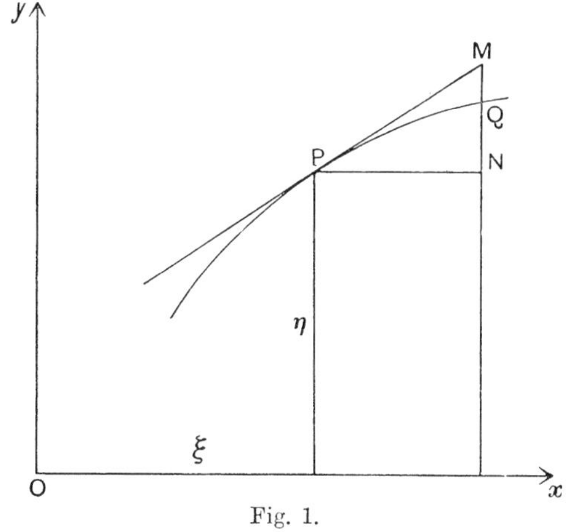

Fig. 1.

Let Q be a neighbouring point of abscissa $\xi + \delta x$, M the point on the ordinate at Q where it cuts the tangent at P, and PN a line through P parallel to Ox, the axis of x. Then

$$QN = \delta y, \quad PN = \delta x.$$

Also the equation of the tangent at P is

$$y - f(\xi) = f'(\xi)\,(x - \xi),$$

so that

$$MN = f'(\xi)\,\delta x = df(x),$$

and therefore

$$(2\text{·}41) \qquad MN = dy.$$

---

* This must be interpreted as in § 2·10 when either $f_x'(x_0,\,y_0)$ or $f_y'(x_0,\,y_0)$ vanishes.

Thus *the differential of y is the increase of the ordinate of a point on the tangent corresponding to an increase of abscissa δx.* This equality holds equally with oblique axes. Of course δx may be an increase of either sign, the increase of the ordinate being reckoned algebraically.

§ 2·50. **Arcs and their differentials.** Any arc of a plane curve which has a continuous tangent is *rectifiable**, and the length of the arc measured from a point P to a point Q (the coordinates being rectangular cartesians) is given by

$$(2\cdot51) \qquad s = \int_{x_0}^{x} \{1 + [f'(x)]^2\}^{\frac{1}{2}}\, dx,$$

$$(2\cdot52) \qquad s = \int_{t_0}^{t} \{[\phi_1'(t)]^2 + [\phi_2'(t)]^2\}^{\frac{1}{2}}\, dt,$$

for curves of the corresponding forms. To avoid ambiguities of sign it has been necessary to assign the direction in which s is to be regarded as increasing along the curve, when measured from a fixed point P. It it usually convenient, though not essential (see § 2·60), to take this direction as the direction of x increasing or t increasing, as has been done in 2·51 and 2·52 above. Taking 2·52 and using differentials we obtain †

$$(2\cdot53) \qquad ds = \{[\phi_1'(t)]^2 + [\phi_2'(t)]^2\}^{\frac{1}{2}}\, dt,$$

$$(2\cdot531) \qquad (ds)^2 = (dx)^2 + (dy)^2 \ddagger.$$

§ 2·60. **Conventions of sign.** We must now make certain conventions to avoid the repeated occurrences of ambiguities of sign. We have three positive directions to assign, namely, the positive axis of x,

---

* See d.l.V.P., Vol. I, pp. 303, 368. For the study of rectifiable curves, more particularly of the *necessary* and sufficient conditions for rectifiability, see d.l.V.P., Vol. I, p. 380; Jordan, Vol. I, p. 99. For the properties of continuous curves in general, see d.l.V.P., Vol. I, p. 374; Jordan, Vol. I, p. 90. The book referred to under the latter title is Jordan's *Cours d'Analyse*, 3rd Ed.

† By the usual rule for differentiating an integral with a *continuous* integrand with respect to the upper limit. The assumption of a continuous tangent is made throughout this section.

‡ Note that 2·531 or its equivalent $ds = \{1 + (dy/dx)^2\}^{\frac{1}{2}}\, dx$ *is not the source of* 2·52 *but a deduction therefrom.* Equation 2·51 or 2·52 is fundamental, for it is a direct deduction from the definition of the length of an arc as the limit of an inscribed polygon, and until 2·51 has been established no meaning can be attached to 2·531.

The reader will of course bear in mind that 2·531 is *not* the same thing as the equation $(\delta s)^2 = (\delta x)^2 + (\delta y)^2$, which is false except for a straight line: 2·531 expresses in the differential notation the fact that, as $\delta x, \delta y \rightarrow 0$,

$$(\delta s)^2 \sim (\delta x)^2 + (\delta y)^2.$$

i.e. the direction of x increasing, the positive axis of y, and the direction, clockwise or counter-clockwise, in which an angle is to be reckoned positive when measured from the positive axis of x. Any two of these three positive directions can be arbitrarily chosen without introducing ambiguities into any of our formulae, but when this has been done, the third positive direction cannot be so chosen if ambiguities are to be avoided. We shall therefore make the following convention.

(A) *The positive axis of y makes an angle $+\frac{1}{2}\pi$ with the positive axis of x.*

With this convention all our formulae remain correct whatever choice is made of the directions in which two of the quantities x, y, and the angle are reckoned positive. We shall, however, in general, suppose that all angles are reckoned positive in a counter-clockwise direction from the positive axis of x. This is convenient though unnecessary. When a concise name is required, we shall denote the positive axes of x and y by Ox and Oy, and similarly the negative axes by Ox' and Oy'.

We have now to assign positive directions along the tangent and normal at any point of a curve, and the direction of s increasing along the curve. We can assign arbitrarily the direction of s increasing along the curve, but once this has been done, no further liberty of choice remains if ambiguities are to be avoided. We make the following conventions.

(B) *The positive direction of the tangent is the direction of a line drawn along the tangent in the direction of s increasing.*

This direction will be spoken of as the direction of the tangent or simply as "the tangent" when no misunderstanding can arise.

(C) *The positive direction of the normal makes an angle $+\frac{1}{2}\pi$ with "the tangent".*

This may be spoken of as the direction of the normal or simply as "the normal". If "the tangent" is the same as Ox, then "the normal"* is the same as Oy.

§ 2·610. **Further differential formulae.** If ψ† is the angle made by the tangent with Ox, we have, in all cases, with the above conventions,

$$\tan \psi = dy/dx = f'(x).$$

It follows at once that

$$\sin \psi = \pm\, dy/ds, \quad \cos \psi = \pm\, dx/ds,$$

---

* We shall not in future emphasise this meaning by inverted commas.

† This use of ψ is constant throughout the rest of Chapters II and III.

and both signs must always be positive*. Therefore

$$(2\text{·}611) \qquad \frac{dx}{ds} = \cos\psi, \qquad \frac{dy}{ds} = \sin\psi,$$

$$(2\text{·}612) \qquad dx = \cos\psi\,ds, \quad dy = \sin\psi\,ds\,\dagger.$$

If l, m ‡ are the direction cosines of the tangent, then

$$l = \cos\psi, \quad m = \sin\psi,$$

$$(2\text{·}613) \qquad \frac{dx}{ds} = l, \qquad \frac{dy}{ds} = m,$$

$$(2\text{·}614) \qquad dx = l\,ds, \quad dy = m\,ds\,\dagger,$$

$$(2\text{·}615) \qquad ds = l\,dx + m\,dy\,\dagger.$$

All these forms express the same facts in different notations, all of utility. The figures illustrate the last sections. The positive directions

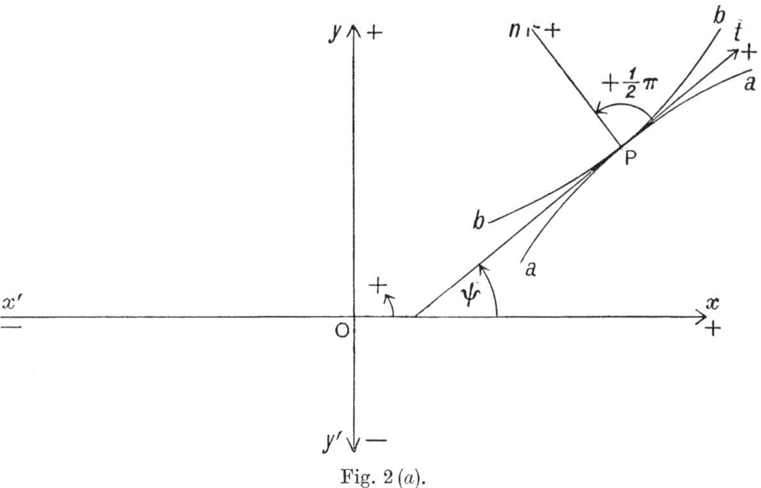

Fig. 2 (a).

of tangent and normal are denoted by Pt and Pn respectively. The two curves shown, aa and bb, have the same tangent and normal at P. It must always be remembered that none of the formulae of this section

----

* The conventions were of course chosen so that this should be so.

† The reader will bear in mind that these differential relations are exact and are not the same as the equations

$$\delta x = \cos\psi\,\delta s, \quad \delta y = \sin\psi\,\delta s,$$

etc. These latter are in fact false unless the curve is a straight line. See equation 2·531, note.

‡ This use of l, m is constant throughout the rest of Chapters II and III.

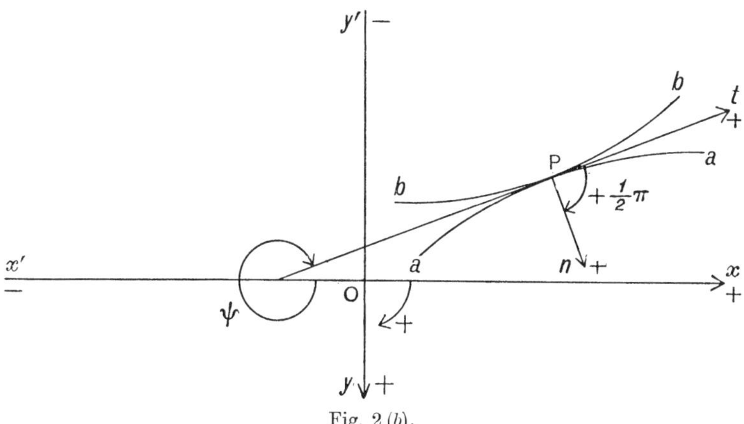

Fig. 2 (*b*).

are affected by the choice of the direction of *s* increasing. We shall return to these formulae when treating of cu·vature.

§ 2·620. **Limiting ratios of arcs, chords, and tangents.** The following theorems deal with the limiting ratios of arcs, chords and tangents.

THEOREM 2·621. *The ratio of the arc and the corresponding chord tends to 1 as the arc tends to 0.*

For
$$(\delta s)^2 = (ds)^2 (1 + o\,(1)),$$
$$(\delta x)^2 + (\delta y)^2 = (dx)^2 (1 + o\,(1)) + (dy)^2 (1 + o\,(1)),$$
$$= \{(dx)^2 + (dy)^2\} (1 + o\,(1)),$$
and therefore
$$(\delta s)^2 \sim (\delta x)^2 + (\delta y)^2,$$
which latter is the square of the chord.

It is easy to show more exactly that, if $f(x)$ has a continuous third differential coefficient (with similar conditions for the other forms of curve),

(2·622)
$$\text{Arc} - \text{Chord} = O\,(\delta s)^3.$$

As a corollary of this it is easy to prove that if μ is the greatest distance between the arc and the chord then

(2·623)
$$\mu = O\,(\delta s)^2.$$

One further step shows that these lengths are effectively of these orders, except for isolated points, unless the curve is a straight line*. These results are frequently useful.

The following result, also useful, is of the same nature as the foregoing. We omit the proof †.

* See Ex. i, 1–3.   † See Ex. i, 5.

THEOREM 2·624. *If PT, QT, are tangents to the curve $y=f(x)$ at P and Q, intersecting at T, then*

$$(2·625) \qquad PT + QT = \delta s + O(\delta s)^3,$$

as $Q \to P$; and, if also $f''(x) \neq 0$ at P,

$$(2·626) \qquad PT - QT = O(\delta s)^2 *.$$

We assume a continuous third differential coefficient. Equations 2·625 and 2·626 hold of course for any form of curve, with a suitable condition to replace $f''(x) \neq 0$. This condition may be stated in the general form that *P must not be a point at which the tangent is stationary* †.

§ 2·70. **Tangents in polar coordinates.** We shall content ourselves with considering a curve of the form $r = f(\theta)$, or one that can be put into that form in the neighbourhood of (r_0, θ_0), the point under consideration.

THEOREM 2·71. *The necessary and sufficient condition that the curve $r = f(\theta)$ may have a tangent at (r_0, θ_0), which is not the radius vector to that point ‡, is that $f'(\theta_0)$ exists.*

The equation of the tangent is then

$$(2·711) \qquad \frac{f(\theta_0)}{r} = \cos(\theta - \theta_0) - \frac{f'(\theta_0)}{f(\theta_0)} \sin(\theta - \theta_0).$$

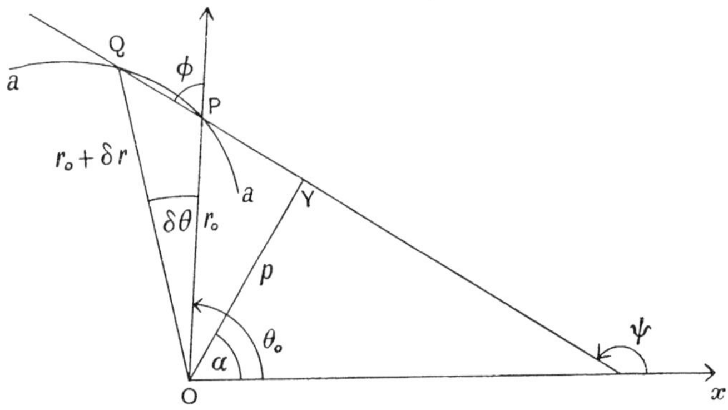

Fig. 3.

If the line QP has a limit as $Q \to P$ which is not the line OP, ϕ (see Figure 3) has a limit which is not 0 or π and so $\sin \phi$ has a non-zero limit. But

$$(r_0 + \delta r) \sin(\phi - \delta\theta) - r_0 \sin \phi = 0,$$

---

* For the case $f''(x) = 0$ see Ex. II, 4.

† See § 3·10 and Ex. II, 4.   ‡ See Theorem 2·11, footnote.

and therefore as $Q \to P$, *i.e.* as δr and $\delta\theta$ tend to zero,

$$\frac{\delta r}{\delta\theta} = (r_0 + \delta r)\left(\frac{\sin \delta\theta}{\delta\theta}\right)\left(\frac{\cos \phi}{\sin \phi}\right)\left(\frac{1}{\cos \delta\theta}\right) + \frac{r_0\,(1 - \cos \delta\theta)}{\delta\theta \cos \delta\theta}\,.$$

The right-hand side tends to a limit, since each term has a limit, and therefore $\delta r/\delta\theta$ has a limit; *i.e.* $f'(\theta_0)$ exists. Conversely, the above reasoning may be reversed, and the theorem is proved.

To find the equation of the tangent when $f'(\theta)$ exists, we proceed as follows. Let $(p,\ a)$ be the polar coordinates of the foot of the perpendicular from O on PQ. Then p and a both tend to finite limits when $Q \to P$. The equation of QPY is

(2·712) $$r \cos (\theta - a) = p\,;$$

and, since P and Q lie on this line,

$$r_0 \cos (\theta_0 - a) = p,$$
$$(r_0 + \delta r) \cos (\theta_0 + \delta\theta - a) = p.$$

Therefore, in the limit, p and a referring to the tangent,

(2·713) $$f(\theta_0) \cos (\theta_0 - a) = p,$$

(2·714) $$f'(\theta_0) \cos (\theta_0 - a) - f(\theta_0) \sin (\theta_0 - a) = 0.$$

To obtain the tangent we find, from 2·712 and 2·713,

$$\frac{f(\theta_0)}{r} = \frac{\cos (\theta - a)}{\cos (\theta_0 - a)} = \frac{\cos (\theta_0 - a + \theta - \theta_0)}{\cos (\theta_0 - a)}$$
$$= \cos (\theta - \theta_0) - \tan (\theta_0 - a) \sin (\theta - \theta_0),$$

and so, by 2·714,

(2·72) $$\frac{f(\theta_0)}{r} = \cos (\theta - \theta_0) - \frac{f'(\theta_0)}{f(\theta_0)} \sin (\theta - \theta_0).$$

§ 2·730. **Conventions of sign and differential formulae in polar coordinates.** We make the following conventions.

The initial line of θ is Ox and θ is measured positive counter-clockwise. When the tangent is continuous, so that the curve is rectifiable, s is so chosen as to increase with θ. The angle ϕ between the tangent and radius vector is reckoned positive counter-clockwise from the positive radius vector* to the positive direction of the tangent. We then have in all cases

(2·731) $$\psi \equiv \theta + \phi, \ (\text{Mod. } 2\pi).$$

---

* The positive radius vector is to mean the direction of an arm drawn from O making an angle $+\theta$ with the initial line Ox. It must be remembered that negative values of r have to be allowed for.

The figures 4 (a) and 4 (b) illustrate this equation.

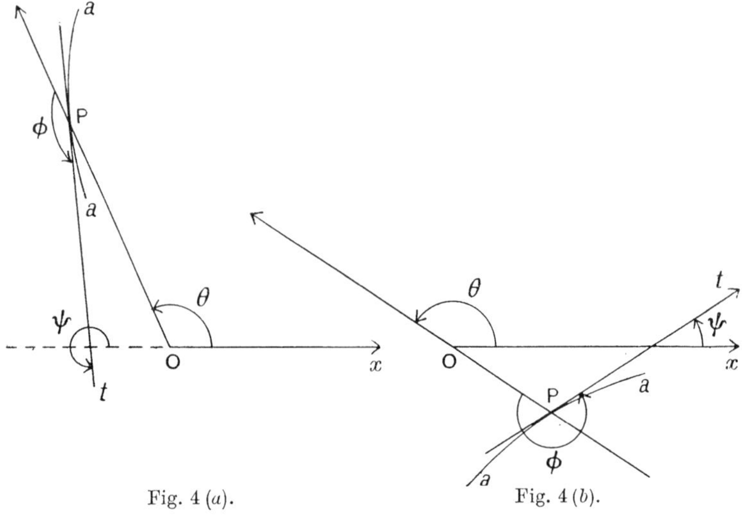

Fig. 4 (a).   Fig. 4 (b).

Referring to Figures 4 and the proof of the last theorem we see that

$$\phi = \tfrac{1}{2}\pi + a - \theta_0,$$

$$\cot \phi = \tan(\theta_0 - a) = \frac{f'(\theta_0)}{f(\theta_0)} = \frac{dr}{r\,d\theta}.$$

This formula is perfectly general, so that we have in all cases the important formula

$$(2\text{·}732) \qquad\qquad \tan \phi = \frac{r\,d\theta}{dr}.$$

To obtain a formula for the differential of the arc, suitable for polar coordinates, we have in all cases

$$x = r \cos \theta, \; y = r \sin \theta,$$

and therefore

$$dx = dr \cos \theta - r\,d\theta \sin \theta,$$
$$dy = dr \sin \theta + r\,d\theta \cos \theta;$$
$$(ds)^2 = (dx)^2 + (dy)^2 = (dr)^2 + r^2 (d\theta)^2,$$
$$(2\text{·}733) \qquad ds = \pm\,\{(dr)^2 + r^2 (d\theta)^2\}^{\frac{1}{2}} = +\,\{r^2 + (dr/d\theta)^2\}^{\frac{1}{2}}\,d\theta,$$
$$\frac{ds}{d\theta} = \left\{r^2 + \left(\frac{dr}{d\theta}\right)^2\right\}^{\frac{1}{2}}.$$

We have also in all cases

$$(2\text{·}734) \qquad\qquad \frac{r\,d\theta}{ds} = \sin \phi^*, \; \frac{dr}{ds} = \cos \phi.$$

____

* Note that ϕ can only be greater than π if r is negative.

§ 2·80. **Concavity and convexity**. Consider a curve given by the equation $y = f(x)$ in rectangular *or oblique* coordinates, and a point P on the curve where $f'(x)$ exists, *i.e.* where the curve has a tangent not parallel to the axis of y. There are two sides to this tangent, which are distinguished by the facts that on one side $y \to +\infty$ along any line parallel to the y-axis, while on the other $y \to -\infty$. We shall call the side on which $y \to +\infty$ the **upper** side, and the side on which $y \to -\infty$ the **lower** side, of this line (the tangent). *If the curve lies entirely on one side of the tangent in the neighbourhood of P, it is said to be* **concave upwards** *or* **convex upwards** *at $P*$ according as it lies on the upper or the lower side of the tangent in the neighbourhood of P.*

THEOREM 2·81. *The curve $y = f(x)$, for which $f(x)$ possesses a continuous second differential coefficient at P, is concave (convex) upwards at P if*

$$f''(x) > 0 \; (< 0).$$

The curve will be concave or convex upwards (see § 2·40) according as $\delta y - dy$ is positive or negative for all sufficiently small values of δx of either sign. But, by Taylor's theorem,

$$(2\cdot811) \qquad \delta y - dy = \tfrac{1}{2}(\delta x)^2 f''(x + \theta \delta x), \; (0 < \theta < 1).$$

As $f''(x)$ is continuous, the theorem follows at once.

A curve $y = f(x)$ is said to be concave (convex) upwards throughout the interval $x_0 \leqslant x \leqslant x_1$, if every point (excluding the end points) of any arc of the curve in this interval lies below (above) the corresponding chord.

This is equivalent to the statement that the curve is concave (convex) upwards throughout the interval, if

$$(2\cdot82) \qquad f(\xi_1) + f(\xi_3) - 2f(\xi_2) > 0 \; (< 0)$$

for all values of ξ_1, ξ_2, ξ_3, such that

$$x_0 \leqslant \xi_1 < \xi_2 < \xi_3 \leqslant x_1.$$

It is easily verified that, if $f''(x)$ is continuous throughout the interval, 2·82 is equivalent to the condition

$$(2\cdot821) \qquad f''(x) > 0 \; (< 0), \; (x_0 \leqslant x \leqslant x_1).$$

It follows at once by comparison of 2·811 and 2·821 that *if such a curve is concave (convex) upwards throughout an interval it is concave (convex) upwards at every point of the interval and conversely.*

---

* Alternatively, convex downwards and concave downwards respectively. It should be remembered that "upwards" is used simply to denote the direction in which $y \to +\infty$. This is because the positive axis of y is usually so drawn, but the word "upwards" must be applied in accordance with the above definition whatever the direction actually chosen for the positive axis of y.

Curves that are concave (convex) upwards throughout an interval are of considerable importance in analysis; their properties can however be discussed without any hypothesis as to the existence of $f''(x)$. Their further treatment is out of place here*.

§ 2·90. **Points of inflexion.** DEFINITION. *A* **point of inflexion** *is a point at which the curve crosses its tangent.*

For the curve $y = f(x)$, with a continuous $f''(x)$, we have

$$\delta y - dy = \tfrac{1}{2} f''(x + \theta \delta x)(\delta x)^2, \quad (0 < \theta < 1).$$

In order that x may be a point of inflexion for such a curve, it is necessary and sufficient that $\delta y - dy$ should change sign with δx, *i.e.* that

$$f''(x + \mu)$$

should change sign with μ. *Hence a point of inflexion is a root of $f''(x) = 0$ at which $f''(x)$ changes sign, and conversely.* A sufficient condition that $f''(x)$ should change sign is that, if $f^{(n)}(x)$ is the first differential coefficient not vanishing with $f''(x)$, n should be odd. For we have

$$\delta y - dy = \frac{(\delta x)^n}{n!} f^{(n)}(x + \theta \delta x), \quad (0 < \theta < 1).$$

Developments of this nature belong more properly to the theory of contact.

It must be remembered, however, that a curve may have a point of inflexion at a point at which the tangent is parallel to the y-axis, and at such a point $f''(x)$ does not exist. To avoid this case of exception, we may say that, for a curve for which one at least of d^2y/dx^2 and d^2x/dy^2 exists and is continuous, the points of inflection are the points at which one at least of d^2y/dx^2 and d^2x/dy^2 vanishes and changes sign.

Consider now the curve

$$x = \phi_1(t), \quad y = \phi_2(t),$$

for which $\phi_1''(t)$, $\phi_2''(t)$ are continuous. The equation of the tangent at t_0 is

$$(x - \phi_1(t_0))\,\phi_2'(t_0) - (y - \phi_2(t_0))\,\phi_1'(t_0) = 0.$$

The perpendicular distance of (x, y), a point on the curve, from the tangent is algebraically proportional to

$$(\phi_1(t) - \phi_1(t_0))\,\phi_2'(t_0) - (\phi_2(t) - \phi_2(t_0))\,\phi_1'(t_0)$$

$$= \tfrac{1}{2}(t - t_0)^2 \begin{vmatrix} \phi_1''(t_0 + \mu) & \phi_2''(t_0 + \mu) \\ \phi_1'(t_0) & \phi_2'(t_0) \end{vmatrix},$$

---

* See d.l.V.P., Vol. I, pp. 285–291.

where $\mu = \theta (t - t_0)$ and $0 < \theta < 1$. The curve therefore crosses the tangent if

$$(2\text{·}901) \qquad \begin{vmatrix} \phi_1{}''(t) & \phi_2{}''(t) \\ \phi_1{}'(t_0) & \phi_2{}'(t_0) \end{vmatrix}$$

vanishes and changes sign at $t = t_0$.

§ 2·910. **Concavity, convexity and points of inflexion in polar coordinates.** The property of being concave or convex *to a point* may be defined in a similar manner to the property of the being concave or convex upwards. If the curve lies entirely on one side of the tangent at P, and Q is a point not lying on the tangent at P, we say that *the curve is concave (convex) at P to the point Q according as the curve near P lies on the same (opposite) side of the tangent at P as the point Q.*

Take the point Q for the origin of polar coordinates, and suppose that the curve is $r = f(\theta)$, and that $f''(\theta)$ is continuous. The equation of the tangent is

$$\frac{f(\theta_0)}{r} = \cos(\theta - \theta_0) - \frac{f'(\theta_0)}{f(\theta_0)} \sin(\theta - \theta_0).$$

The curve will be concave (convex) to the origin if

$$\frac{f(\theta_0)}{f(\theta)} - \cos(\theta - \theta_0) + \frac{f'(\theta_0)}{f(\theta)} \sin(\theta - \theta_0) > 0 \; (< 0),$$

when $|\theta - \theta_0|$ is small. This condition reduces to

$$(f(\theta_0))^2 + 2(f'(\theta_0))^2 - f(\theta_0) f''(\theta_0) > 0 \; (< 0).$$

In the same way the condition for a point of inflexion for such a curve is that

$$(f(\theta))^2 + 2(f'(\theta))^2 - f(\theta) f''(\theta)$$

should vanish and change sign at $\theta = \theta_0$.

Much of what proceeds takes a simpler form if the curve is expressed as

$$u = f(\theta),$$

where $u = 1/r$. The last expression in the condition for an inflexion is then replaced by

$$f(\theta) + f''(\theta).$$

EXAMPLES I

(1) Prove that the μ of § 2·620 satisfies the relation

$$\mu \sim \tfrac{1}{8} f''(x_0)(\delta x)^2$$

for the curve $y = f(x)$, and deduce that $\mu = o(\delta x)^2$ implies that the curve is a straight line.

[At the point of greatest distance between arc and chord, the tangent must be parallel to the chord. Hence take the tangent and normal as axes and consider the intersections of $y=f(x)$ $(f(0)=f'(0)=0)$ and $y=\mu.$]

(2) Taking $x=f(t)$, $y=g(t)$ for the curve and writing

$$\Omega^2=(f'(t))^2+(g'(t))^2,$$

prove that

$$\text{arc}-\text{chord} \sim \frac{(\delta t)^3}{24\,\Omega^3}\,(g''f'-f''g')^2$$

$$=(\delta s)^3/24\rho^2.$$ (see 3·132)

Deduce that

$$\text{arc}-\text{chord}=o\,(\delta s)^3$$

implies that the curve is a straight line.

[We have

$$\delta\,(\text{arc})=\int_0^{\delta t}\{(f'(t))^2+(g(t))^2\}^{\frac{1}{2}}\,dt,$$

$$\delta\,(\text{chord})=\{(f(t+\delta t)-f(t))^2+(g(t+\delta t)-g(t))^2\}^{\frac{1}{2}}.$$

Expand by Taylor's theorem in powers of δt as far as $(\delta t)^3$, and the result follows.   If $(\text{arc}-\text{chord})=o\,(\delta t)^3$, then $f''/f'=g''/g'$.   Integrating, $f'=Ag'$, $f=Ag+B$, which defines a straight line.]

(3) Deduce $\mu=O\,(\delta s)^2$ directly from

$$\text{arc}-\text{chord}=O\,(\delta s)^3.$$

[Establish and use the fact that

$$\mu^2 < \tfrac{1}{4}\,\{(\text{arc})^2-(\text{chord})^2\}.]$$

(4) Using polar coordinates, prove that, if $f'(\theta)$ is continuous at $\theta=\theta_0$, the tangent is the limit of the chord Q_1Q_2 as Q_1, $Q_2 \rightarrow P\,(r_0,\,\theta_0)$.

[Combine the proofs of 2·71 and 2·21.]

(5) With the notation of Theorem 2·624 prove that, in order that $PT/QT \rightarrow 1$, it is sufficient that $f''(x)$ should be continuous and not zero at P.

[Take the tangent and normal as axes, and the curve as

$$y=f(x),\quad (f(0)=f'(0)=0).$$

The tangent at x is given by 2·11. Find the coordinates of T. Theorem 2·624 itself may be established similarly. Note that the condition $f''(x)\neq0$ is invariant, § 1·40.]

(6) Three neighbouring tangents are drawn to a curve. If δs_1, δs_2 be the lengths of the arcs between the points of contact taken in order along the curve, $\delta\psi$ the angle between the extreme tangents, and Δ the area of the enclosed triangle, then

$$\Delta \sim \tfrac{1}{8}\delta s_1\,\delta s_2\,\delta\psi.$$

It is assumed that the points of contact tend to a point at which $f''(x)$ is continuous and not zero.

[The lengths of the two sides including the angle $\delta\psi$ are asymptotically equivalent to $\tfrac{1}{2}\delta s_1$ and $\tfrac{1}{2}\delta s_2$ respectively. This follows at once from Exs. 3 and 5.]

(7) Prove geometrically that, if PT the tangent at P exists, then the chord Q_1Q_2 tends to the tangent at P as Q_1, $Q_2 \to P$ from opposite sides. Show where the proof fails when Q_1, $Q_2 \to P$ in any manner.

Show that if PQ_1, PQ_2 are equal and opposite arcs, and $f'''(x)$ exists at P, the chord Q_1Q_2 represents the tangent at P with an error $O(\delta s)^2$: and that some condition of equality is essential.

[The angle between PT and Q_1Q_2 is less than the angle between PT and PQ_1. Hence Q_1Q_2 tends to parallelism with PT, etc.]

(8) Define a tangent of the twisted curve

$$x = \phi_1(t), \ y = \phi_2(t), \ z = \phi_3(t),$$

and obtain its equations in the forms

$$\frac{x - x_0}{\phi_1'} = \frac{y - y_0}{\phi_2'} = \frac{z - z_0}{\phi_3'},$$

$$\frac{x - x_0}{dx} = \frac{y - y_0}{dy} = \frac{z - z_0}{dz}.$$

If l_1, m_1, n_1 are the direction cosines of the tangent, prove that

$$dx = l_1 ds, \ dy = m_1 ds, \ dz = n_1 ds.$$

[d.l.V.P., Vol. I, pp. 325–331.]

(9) The osculating plane of a twisted curve being defined as the limit of the plane PQ_1Q_2 when Q_1 and Q_2 tend to P along the curve, obtain its equation in the forms

$$\begin{vmatrix} x - x_0 & y - y_0 & z - z_0 \\ \phi_1' & \phi_2' & \phi_3' \\ \phi_1'' & \phi_2'' & \phi_3'' \end{vmatrix} = 0, \qquad \begin{vmatrix} x - x_0 & y - y_0 & z - z_0 \\ dx & dy & dz \\ d^2x & d^2y & d^2z \end{vmatrix} = 0.$$

Prove that the osculating plane is the limit of

(a) a plane through the tangent at P and the chord PQ,

(β) a plane through the tangent at P and parallel to the tangent at Q. Alternatively, show that the angle between the normal to the osculating plane at P and the tangent at Q is $\tfrac{1}{2}\pi + o(\delta s)$, if ϕ_1'' etc. exist at P, and is $\tfrac{1}{2}\pi + O(\delta s)^2$ if ϕ_1''' etc. exist at P.

[d.l.V.P., Vol. I, pp. 335–337.]

(10) If PQ_1, PQ_2 are equal and opposite arcs δs, prove that under suitable conditions the plane PQ_1Q_2 represents the osculating plane at P with an error $O(\delta s)^2$, but that if the arcs are not equal the error will usually be $O(\delta s)$.

(11) The line of intersection of the osculating plane at P with the osculating plane at Q tends to the tangent at P as $Q \to P$.

[d.l.V.P., Vol. I, p. 346: or geometrically, using Ex. 10.]

CHAPTER III

THE CURVATURE OF PLANE CURVES

§ 3·10. **Curvature.** The idea of curvature is introduced to afford a measure of the rate at which the tangent is turning as the point of contact moves along the curve. Suppose that PQ are two points on any rectifiable curve which has a tangent at every point of the arc PQ, δs the length of the arc PQ, and $\delta\psi$ the angle between the tangents at P and Q. Then $\delta\psi/\delta s$ is called *the mean curvature of the arc PQ*, and

$$\operatorname*{Lt}_{Q \to P} \frac{\delta\psi}{\delta s},$$

if it exists, is called *the curvature at P*. If this limit is denoted by $1/\rho$, ρ is called *the radius of curvature at P*.*

We shall find in practice that it is necessary to attach a *sign* to the mean curvature, curvature, and radius of curvature. Consider the case of two equal circles touching externally at P. The mean curvatures of

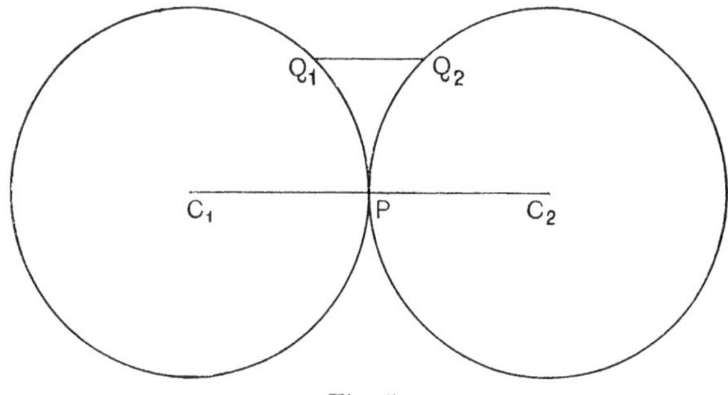

Fig. 5.

the equal arcs PQ_1, PQ_2 are, as at present defined, equal, though the tangents turn in opposite directions as the points of contact go from P to Q_1 and Q_2. To take account of this we introduce the following definitions.

---

* If the curvature at P is zero, there is no radius of curvature or "an infinite radius of curvature" according to choice.

DEFINITIONS. *With the conventions of* § 2·60, **the mean curvature of the arc** PQ *is defined as*

$$(\psi_Q - \psi_P)/(s_Q - s_P),$$

the arc s being measured from any convenient point on the curve, and the **curvature** *at P is defined as*

$$\underset{Q \to P}{\mathrm{Lt}}\ (\psi_Q - \psi_P)/(s_Q - s_P),$$

if this limit exists.

The radius of curvature at P *is the reciprocal of the curvature, and is usually denoted by* ρ.

The centre of curvature at P *is a point C on the normal at P such that* $PC = \rho$*.

The circle of curvature at P *is a circle with centre C and radius* $PC\,(=\rho)$.

We find in all cases that the coordinates $(X,\ Y)$ of C satisfy

(3·11) $$X = x - \rho \sin \psi, \quad Y = y + \rho \cos \psi;$$

or, if $(l',\ m')$ are the direction cosines of the positive normal,

(3·12) $$X = x + l'\rho, \quad Y = y + m'\rho.$$

THEOREM 3·13. *The necessary and sufficient condition that the curve $y = f(x)$ should have a definite curvature at any point x, where the tangent is not parallel to Oy, is that $f''(x)$ should exist. The curvature $1/\rho$ is then given by*

(3·131) $$\frac{1}{\rho} = \frac{f''(x)}{\{1 + (f'(x))^2\}^{\frac{3}{2}}}.$$

(1) The condition is necessary. For in order that ψ_P and ψ_Q may exist, it is necessary that the curve should have a tangent in the neighbourhood of P; and in order that $\mathrm{Lt}\,(\psi_Q - \psi_P)/(s_Q - s_P)$ may exist it is necessary that $\psi_Q \to \psi_P$, *i.e.* that ψ should be continuous at P. As

(3·1311) $$\tan \psi = f'(x), \quad (\psi \neq \tfrac{1}{2}\pi),$$

$f'(x)$ must be continuous at P.

If Q is the point $(x + \delta x,\ y + \delta y)$,

$$\delta s = \int_x^{x + \delta x} \{1 + (f'(\xi))^2\}^{\frac{1}{2}}\, d\xi.$$

Since $f'(\xi)$ is continuous at $\xi = x$,

$$\{1 + (f'(\xi))^2\}^{\frac{1}{2}} = \{1 + (f'(x))^2\}^{\frac{1}{2}} + o\,(1),$$

$$\delta s = \delta x\,\{1 + (f'(x))^2\}^{\frac{1}{2}} + o\,(\delta x).$$

---

* Of course with due regard to the sign of ρ and the positive direction of the normal Pn. If ρ is positive C lies on the positive normal Pn; if ρ is negative C lies on Pn'.

Therefore
$$\underset{\delta x \to 0}{\text{Lt}} \frac{\delta x}{\delta s} = \{1 + (f'(x))^2\}^{-\frac{1}{2}} \neq 0.$$

Also
$$\delta \psi / \delta s = (\delta \psi / \delta x) \times (\delta x / \delta s),$$

and therefore, as $\delta\psi/\delta s$ has a limit, and $\delta x/\delta s$ a non-zero limit, as $\delta x \to 0$, $\delta\psi/\delta x$ must have a limit. But applying the mean value theorem to 3·1311 we have

$$\delta\psi \sec^2(\psi + \theta' \delta\psi) = f'(x + \delta x) - f'(x), \quad (0 < \theta' < 1),$$

$$\frac{\delta\psi}{\delta x} = \frac{f'(x + \delta x) - f'(x)}{\delta x} \cos^2(\psi + \theta' \delta\psi).$$

But $\cos^2(\psi + \theta' \delta\psi)$ has the non-zero limit $(1 + \{f'(x)\}^2)^{-1}$, so that

$$\{f'(x + \delta x) - f'(x)\} / \delta x$$

must have a limit, *i.e.* $f''(x)$ must exist.

(2) Conversely, the above reasoning may be reversed, and if $f''(x)$ exists, then

$$\underset{\delta x \to 0}{\text{Lt}} \frac{\delta\psi}{\delta x} = \frac{f''(x)}{1 + \{f'(x)\}^2},$$

$$\underset{\delta x \to 0}{\text{Lt}} \frac{\delta x}{\delta s} = (1 + \{f'(x)\}^2)^{-\frac{1}{2}};$$

so that
$$\underset{\delta s \to 0}{\text{Lt}} \frac{\delta\psi}{\delta s} = \underset{\delta x \to 0}{\text{Lt}} \frac{\delta\psi}{\delta x} \times \underset{\delta x \to 0}{\text{Lt}} \frac{\delta x}{\delta s};$$

and finally
$$\frac{1}{\rho} = \frac{f''(x)}{(1 + \{f'(x)\}^2)^{\frac{3}{2}}}.*$$

At points at which $d\psi/ds = 0$, the curve is said to have *a stationary tangent*. It is easy to see that, if x_0 is such a point, $f''(x_0) = 0$ provided $f''(x_0)$ exists. At all points of inflexion, therefore, at which $f''(x)$ exists, the curve has a stationary tangent, but the converse is not true, for at a point of stationary tangent the curve may not cross its tangent. The further consideration of such points belongs more properly to the theory of singular points.

Other formulae for the curvature $1/\rho$ are

$$(3\cdot132) \qquad \frac{1}{\rho} = \frac{\phi_1'\phi_2'' - \phi_2'\phi_1''}{(\phi_1'^2 + \phi_2'^2)^{\frac{3}{2}}},$$

when the curve is $x = \phi_1(t)$, $y = \phi_2(t)$; and

$$(3\cdot133) \qquad \frac{1}{\rho} = -\frac{F_{xx}'' F_y'^2 - 2F_{xy}'' F_x' F_y' + F_{yy}'' F_x'^2}{(F_x'^2 + F_y'^2)^{\frac{3}{2}}},$$

---

* If $Q \to P$ from the right (left) then $f''(x)$ need not exist, but it is necessary and sufficient that $f'(x)$ should have a differential coefficient on the right (left).

when the curve is $F(x, y) = 0$. These are direct deductions from $(3{\cdot}131)$ and the theorems of the differential calculus. The proofs may be left to the reader.

§ 3·20. **Properties of the centre of curvature.** The centre of curvature C has a large number of properties many of which give rise to alternative definitions. These are contained in the following theorems.

THEOREM 3·21. *A curve is always concave to its centre of curvature.* The proof of this theorem is left to the reader.

THEOREM 3·22. *If the normal at Q cuts the normal at P in Γ, and C the centre of curvature at P exists, then $\Gamma \to C$ as $Q \to P$.*

Take as axes of coordinates the tangent and normal to the curve at P. By § 1·40 this does not affect the generality of the argument. Then, by Theorem 1·52, the equation of a sufficiently small arc of the curve containing P can be put in the form

$$y = f(x), \quad (\,|\,x\,| < \delta),$$

where $f'(0) = 0$ and $f''(0)$ exists and is not zero*. The coordinates of C are $(0, \rho)$ or $(0, 1/f''(0))$. The equation of the normal at $Q(\xi, f(\xi))$ is

$$f'(\xi)\,(y - f(\xi)) + (x - \xi) = 0,$$

which cuts $x = 0$ where

$$y = f(\xi) + \xi/f'(\xi).$$

This is the ordinate of Γ. As $Q \to P$, $\xi \to 0$, $f(\xi) \to 0$, and

$$\xi/f'(\xi) = (\xi - 0)/(f'(\xi) - f'(0)) \to 1/f''(0)$$

by definition, since $f''(0)$ exists *and is not zero*. Hence $\Gamma \to C$.

THEOREM 3·23. (A) *If the normal at Q_1 cuts the normal at Q_2 in Γ, and C the centre of curvature at P exists, then $\Gamma \to C$ as $Q_1 \to P$ and $Q_2 \to P$, if Q_1 and Q_2 are always on opposite sides of P.*

(B) *But if $Q_1 \to P$ and $Q_2 \to P$ in any manner, Γ need not tend to C unless ρ is continuous at P.*

(A) Let $Q_1\Gamma$ cut PC in Γ_1, $Q_2\Gamma$ cut PC in Γ_2; then, if Q_1 and Q_2 are always on opposite sides of P, Γ cannot lie on $Q_1\Gamma_1$ and $Q_2\Gamma_2$ or on both $Q_1\Gamma_1$ and $Q_2\Gamma_2$ produced; it can only lie on one of them and on the other produced, as in the figure. It is true that Γ, Γ_1, Γ_2 can all coincide, but this will not affect the succeeding argument. As $Q_1\Gamma Q_2 \to 0$, $\Gamma_1\Gamma\Gamma_2$ is

---

* The existence of C implies that C is a finite point according to our usage.

ultimately greater than $\frac{1}{2}\pi$, and therefore Γ lies inside a circle on $\Gamma_1\Gamma_2$ as diameter. But $\Gamma_1 \to C$, $\Gamma_2 \to C$ as $Q_1 \to P$, $Q_2 \to P$; it therefore follows that $\Gamma \to C$ as Q_1 and Q_2 tend to P from opposite sides.

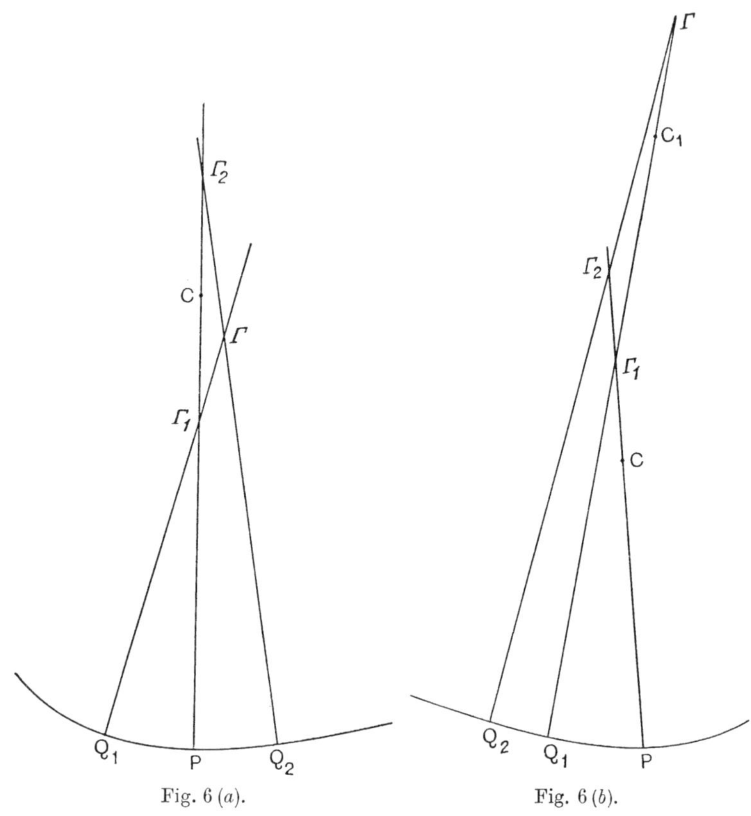

Fig. 6 (a).   Fig. 6 (b).

(B) The preceding argument breaks down here. But if ρ is continuous at P, there must exist a centre of curvature C_1 at Q_1 if Q_1 is sufficiently near P. By the preceding reasoning, if Q_1 is the middle point of the three Q_2, Q_1, P, then Γ, Γ_1, $\Gamma_2 \to C_1$ as $Q_2 \to Q_1$, $P \to Q_1$, Q_1 remaining fixed. In other words, given ϵ we can find a number δ_1 such that $\Gamma C_1 < \epsilon$ provided only $Q_2 Q_1 < \delta_1$, $PQ_1 < \delta_1$. But since ρ is continuous at P, $C_1 \to C$ as $Q_1 \to P$, or in other words there exists a number δ_2 such that $C_1 C < \epsilon$ if only $Q_1 P < \delta_2$. Hence there exists a number δ (the lesser of δ_1 and δ_2) such that $C\Gamma < 2\epsilon$ if only $Q_1 P < \delta$, $Q_2 P < \delta$. In other words $\Gamma \to C$ as required.

On the other hand it is possible to construct cases in which, if ρ is not continuous at P, Γ does not tend to C as Q_1, $Q_2 \to P$ in a suitable manner.

This theorem is an example of the general principle explained in Note A to § 1·30 of the Introduction.

THEOREM 3·24. *If a circle whose centre is O cut* a regular arc of any curve in two points Q_1, Q_2, then there exists a point P on the curve between† Q_1 and Q_2 such that OP is a normal to the curve.*

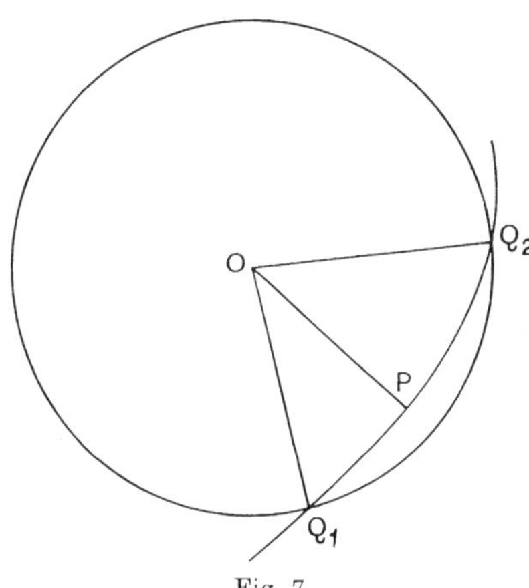

Fig. 7.

By a "regular" arc we mean an arc along which the curve can be put in the general form
$$x = \phi_1(t), \quad y = \phi_2(t),$$
where $\phi_1'(t)$, $\phi_2'(t)$ exist, and are not both zero, at Q_1, Q_2 and all points of the arc between.

We have, if O is the point (a, b), and P any point on the arc,
$$OP^2 = (a - \phi_1(t))^2 + (b - \phi_2(t))^2;$$
therefore OP^2 is a continuous function of t while P lies on the arc Q_1Q_2, possessing everywhere a differential coefficient
$$- 2 \{(a - \phi_1) \phi_1' + (b - \phi_2) \phi_2'\},$$
and taking the same value (R^2 say) both at Q_1 and Q_2. Therefore by

---

* Nothing prevents the curve and circle from *touching* at one or both of Q_1 and Q_2.

† *I.e.* a value of t such that $t_1 < t < t_2$ if t_1 and t_2 are the parameters of the points Q_1, Q_2.

Rolle's Theorem, there exists a point P between Q_1 and Q_2* such that $\frac{d}{dt}(OP^2) = 0$, *i.e.* such that

$$(a - \phi_1)\,\phi_1' + (b - \phi_2)\,\phi_2' = 0.$$

This is satisfied at any point at which ϕ_1' and ϕ_2' both vanish. But if as we have presupposed they do not, it is the condition that OP is normal to the curve at the point P as required.

We can now prove at once two more theorems embodying possible definitions of C.

THEOREM 3·25. *If a circle centre O be drawn touching the curve at P and cutting it at Q, and if C the centre of curvature at P exists, then $O \to C$ as $Q \to P$.*

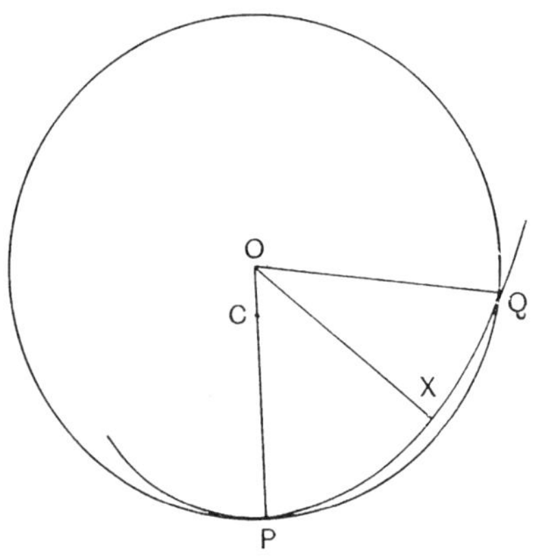

Fig. 8.

Since C exists, the conditions of Theorem 3·24 are satisfied when Q is sufficiently close to P. Therefore there exists a point X between Q and P such that OX is normal to the curve. As $Q \to P$, $X \to P$, and as $X \to P$, $O \to C$, by Theorem 3·22, as was to be proved.

In a similar manner we prove

THEOREM 3·26. *If a circle centre O be drawn cutting the curve at three points Q_1, Q_2, Q_3, and if ρ is continuous at P, then $O \to C$ as Q_1, Q_2, $Q_3 \to P$†.*

* *I.e.* a value of t such that $t_1 < t < t_2$ if t_1 and t_2 are the parameters of the points Q_1, Q_2.

† Note the following enunciation which does not involve the hypothesis of continuity :—

If a circle centre O be drawn cutting the curve at three points Q_1, P, Q_2 and C, the centre of curvature at P, exists, then $O \to C$ as Q_1, $Q_2 \to P$ from opposite sides.

THEOREM 3·27. *The osculating circle, when it exists, is the same as the circle of curvature.*

This is a particular case of theorems on contact. See § 4·40.

Each of the foregoing theorems embodies a possible definition of C the centre of curvature at a point P, but it is clear that some are more general than others. Thus any of the definitions embodied in Theorems 3·22, 3·23 (A), 3·25, are as general as our definition, and accordingly may replace it. The others are not so general and so are not suitable. Conversely, it may be shown that if any of these alternative definitions be adopted, then, at any point at which the curve has a curvature $1/\rho$ by the new definition,

$$\operatorname*{Lt}_{\delta s \to 0} \frac{\delta\psi}{\delta s} = \frac{1}{\rho}.$$

It follows that these definitions are completely equivalent. As a specimen of the theorems to be proved we take the following:

If the normal at Q cuts the normal at P in Γ, and if Γ tends to a finite limit C as $Q \to P$, then

$$\operatorname*{Lt}_{\delta s \to 0} \frac{\delta\psi}{\delta s} = \frac{1}{PC},$$

with the proper conventions of sign.

This is the converse of Theorem 3·22, and the proof is simply the proof of that theorem turned backwards, with obvious changes.

There are a variety of other circles which tend to the circle of curvature at P as a limit. The following are instances.

A circle touching the curve at P, and the tangent at a neighbouring point Q.*

A circle touching the tangent at P and the tangents at neighbouring points Q_1, Q_2†.*

A circle passing through P and touching two tangents at neighbouring points Q_1, Q_2‡.

A circle touching the tangent at P, and passing through two neighbouring points Q_1, Q_2‡.

In each case P may be replaced by a point Q_3 which tends to P. Direct proofs of all these theorems are not difficult (with the obvious

---

* *I.e.*, the circle touches Qt, the tangent to the curve at Q, but touches Qt not necessarily at Q.

† There are four circles touching these three tangents, of which three have as limit the point P.

‡ There are two such circles in each case, both of which have the circle of curvature at P as limit.

assumptions). They depend on a combination of Theorems 3·25 and 3·26 with well-known properties of triangles.

§ 3·30. **The closeness of approximations to the circle of curvature.** It is often useful to know the order of the error involved in replacing the circle of curvature by one of the circles which has the circle of curvature for a limit, for instance the circle through P and two neighbouring points Q_1, Q_2.

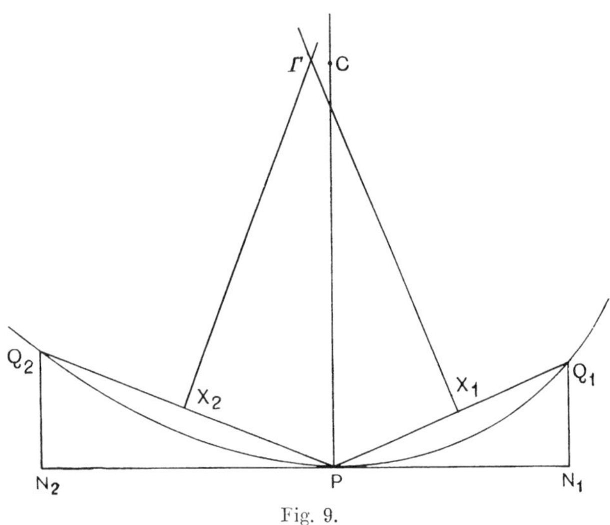

Fig. 9.

Taking the tangent and normal at P for axes, the curve is

$$y = f(x), \quad (f(0) = f'(0) = 0, \quad f''(0) \neq 0);$$

C is the point $(0, 1/f''(0))$, and Γ the intersection of the perpendicular bisectors of PQ_1, PQ_2. Then Γ lies on the two lines

$$(3·31) \qquad \begin{cases} \xi_1 x + \eta_1 y - \tfrac{1}{2}(\xi_1{}^2 + \eta_1{}^2) = 0, \\ \xi_2 x + \eta_2 y - \tfrac{1}{2}(\xi_2{}^2 + \eta_2{}^2) = 0, \end{cases}$$

where Q_1 is (ξ_1, η_1) and Q_2 is (ξ_2, η_2). If $f^{iv}(0)$ exists, we have

$$\eta_1 = \tfrac{1}{2}\xi_1{}^2 f''(0) + \tfrac{1}{6}\xi_1{}^3 f'''(0) + O(\xi_1{}^4),$$

and therefore the equations (3·31) reduce to

$$x + \{\tfrac{1}{2}\xi_1 f''(0) + \tfrac{1}{6}\xi_1{}^2 f'''(0) + O(\xi_1{}^3)\} y - \tfrac{1}{2}\xi_1(1 + O(\xi_1{}^2)) = 0,$$

$$x + \{\tfrac{1}{2}\xi_2 f''(0) + \tfrac{1}{6}\xi_2{}^2 f'''(0) + O(\xi_2{}^3)\} y - \tfrac{1}{2}\xi_2(1 + O(\xi_2{}^2)) = 0.$$

Solving, we have

$$\frac{x}{O\left(\xi^{3}\right)}=\frac{y}{\frac{1}{2}\left(\xi_{2}-\xi_{1}\right)+O\left(\xi^{3}\right)}=\frac{1}{\frac{1}{2}\left(\xi_{2}-\xi_{1}\right)f''(0)+\frac{1}{6}\left(\xi_{2}^{2}-\xi_{1}^{2}\right)f'''(0)+O\left(\xi^{3}\right)},$$

where ξ denotes the numerically greater of ξ_{1} and ξ_{2}.

It follows that in general the ordinates of C and Γ differ by a length of order ξ, but that if $\xi_{2}=-\xi_{1}$ then

$$(3\text{·}32)\qquad\qquad x=O\left(\xi^{2}\right),\quad y=\frac{1}{f''(0)}+O\left(\xi^{2}\right).$$

We have therefore proved that *if the arcs PQ_{1}, PQ_{2} are equal and opposite (say $\pm\,\delta s$), or more generally if $PQ_{1}=\delta s$, $PQ_{2}=-\,\delta s+O\left(\delta s\right)^{2}$, then*

$$(3\text{·}33)\qquad\qquad\qquad \Gamma C=O\left(\delta s\right)^{2}.$$

Since both circles pass through P, their radii also differ by $O\left(\delta s\right)^{2}$, and in fact *the circle $Q_{2}PQ_{1}$ represents the circle of curvature at P with an error $O\left(\delta s\right)^{2}$.*

This fact is important if we wish to apply geometrical reasoning to approximate figures. If, for instance, we are going to argue about C and C_{1} (the centre of curvature at Q_{1}), representing them in the foregoing manner by Γ and Γ_{1}, $\Gamma\Gamma_{1}$ or CC_{1} will itself be a small quantity of the first order and, unless ΓC and $\Gamma_{1}C_{1}$ are of higher order than the first, no such argument can possibly be legitimate. It is essential to take the precaution indicated by the preceding discussion*. When this has been done we have for instance

$$\Gamma\Gamma_{1}/CC_{1}=1+O\left(\delta s\right),$$

whereas in the other case it is by no means evident even that

$$\text{Lt } \Gamma\Gamma_{1}/CC_{1}=1.$$

In a similar way it may be shown that, if K is the intersection of normals at Q_{1} and Q_{2},

$$(3\text{·}34)\qquad\qquad\qquad KC=O\left(\delta s\right)^{2}$$

if and only if PQ_{1} and PQ_{2} are, with a possible error $O\left(\delta s\right)^{2}$, equal and opposite arcs.

---

* Similar arguments must be applied in the theory of twisted curves when approximate geometrical figures are used. The use of such figures can only be legitimate when the errors in the approximate representations are of the second or higher order. This point appears to be often overlooked.

§ 3·40. **Newton's Method.** A formula that is often useful is contained in the following theorem.

THEOREM 3·41. *If the tangent and normal at P are taken as axes of x and y, and ρ exists, then*

$$(3\cdot42) \qquad\qquad \rho = \operatorname*{Lt}_{x \to 0} \frac{x^2}{2y}.$$

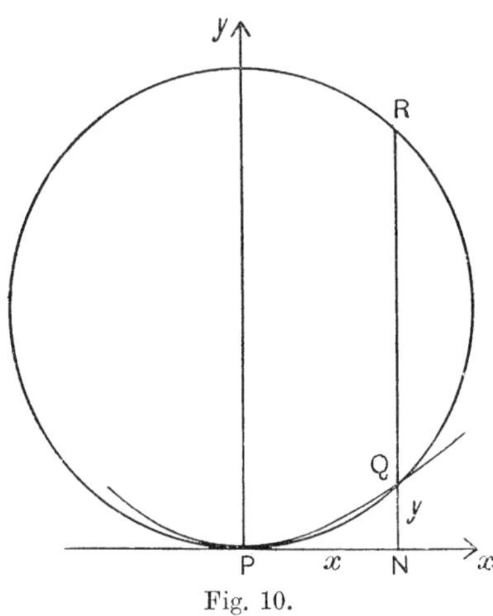

Fig. 10.

Let Q be the point (x, y). Draw a circle to touch the curve at P and pass through Q; then, by Theorem 3·25, this circle tends to the circle of curvature as $Q \to P$, *i.e.* as $x \to 0$. If R is the radius of this circle, $RN = 2R - y$. We have also

$$NP^2 = NQ \,.\, NR,$$

so that
$$x^2/y = 2R - y \to 2\rho,$$

which proves the theorem.

§ 3·50. **The Differentials** dl, dm, **etc.** Theorem 3·22 has an important consequence. The equation of the normal at (ξ, η) may be written

$$(3\cdot501) \qquad\qquad (x - \xi)\, l + (y - \eta)\, m = 0,$$

where (l, m) are as usual the direction cosines of the tangent at (ξ, η). $C(X, Y)$ lies on this line, so that

$$(3\cdot502) \qquad\qquad (X - \xi)\, l + (Y - \eta)\, m = 0.$$

But C is the limit of intersections of neighbouring normals, and therefore* (X, Y) also satisfy the equation

$$(3\text{·}503) \qquad (X - \xi)\, dl + (Y - \eta)\, dm = l\, d\xi + m\, d\eta = ds.$$

But we have also $l^2 + m^2 = 1$, and therefore

$$l\, dl + m\, dm = 0.$$

Hence dl and dm satisfy

$$(3\text{·}504) \qquad \begin{cases} (Y - \eta)\, dl - (X - \xi)\, dm = 0, \\ (X - \xi)\, dl + (Y - \eta)\, dm = ds\,; \end{cases}$$

which may also be written

$$(3\text{·}505) \qquad \begin{cases} m'\, dl - l'\, dm = 0, \\ l'\, dl + m'\, dm = ds/\rho, \end{cases}$$

where (l', m') are the direction cosines of the positive normal.

We have therefore

$$(3\text{·}51) \qquad \frac{dl}{ds} = \frac{l'}{\rho}, \quad \frac{dm}{ds} = \frac{m'}{\rho},$$

$$(3\text{·}52) \qquad \frac{1}{\rho^2} = \left(\frac{dl}{ds}\right)^2 + \left(\frac{dm}{ds}\right)^2.$$

Remembering that $l = dx/ds$, $m = dy/ds$, we have

$$(3\text{·}53) \qquad \frac{1}{\rho^2} = \left(\frac{d^2x}{ds^2}\right)^2 + \left(\frac{d^2y}{ds^2}\right)^2.$$

With our conventions†, l' and m' are connected with l and m by the
relations

$$l' = m, \quad m' = -l.$$

---

* The full reasoning here is as follows. (X, Y) is the limit of the intersection of

$$(x - \xi)\, l + (y - \eta)\, m = 0$$

and $\qquad (x - \xi - \delta\xi)(l + \delta l) + (y - \eta - \delta\eta)(m + \delta m) = 0,$

i.e. of $\qquad (x - \xi)\, l + (y - \eta)\, m = 0$

and $\qquad (x - \xi)\, \delta l + (y - \eta)\, \delta m = \{(l + \delta l)\, \delta\xi + (m + \delta m)\, \delta\eta\}.$

Now when $f''(x)$ exists $d\psi$ exists, and therefore dl and dm exist. Therefore (X, Y)
satisfy
$$(x - \xi)\, dl + (y - \eta)\, dm = ds,$$
which is the limiting form of the second equation.

† In general (l', m') satisfy

$$l'^2 + m'^2 = 1, \quad ll' + mm' = 0$$

(condition of perpendicularity), and therefore

$$l' = \lambda m, \quad m' = -\lambda l,$$

where $\lambda = \pm 1$. $\lambda = 1$ when the conventions are chosen so that, the positive tangent
coinciding with Ox, the positive normal coincides with Oy; $\lambda = -1$ when, in like
case, the positive normal coincides with Oy'.

Therefore

$$(3\text{·}54) \qquad \frac{dl'}{ds} = -\frac{l}{\rho}, \quad \frac{dm'}{ds} = -\frac{m}{\rho}.$$

It may be noted that we have nowhere assumed the continuity of $f''(x)$ or its equivalent. Formulae 3·51, 3·54 are the two-dimensional analogues of Frenet's Formulae.

A more general formula for ρ, available for any parametric representation of the curve, may be obtained as follows from 3·52:

$$\frac{(ds)^2}{\rho^2} = (dl)^2 + (dm)^2$$

$$= \left(d\left(\frac{dx}{ds}\right)\right)^2 + \left(d\left(\frac{dy}{ds}\right)\right)^2$$

$$= \frac{(ds\,d^2x - dx\,d^2s)^2 + (ds\,d^2y - dy\,d^2s)^2}{(ds)^4},$$

which reduces to

$$(3\text{·}55) \qquad \frac{1}{\rho^2} = \frac{(d^2x)^2 + (d^2y)^2 - (d^2s)^2}{(ds)^4}.$$

§ 3·60. **An alternative treatment using moving axes.** The theorems of § 3·50 can, alternatively, be obtained by the use of moving axes. This method has the advantage that it can be immediately generalized to provide a simple treatment of the corresponding problems for twisted curves.

The simplification obtained by this treatment is essentially due to the fact that the curve is at each point P referred to its tangent Pt and normal Pn as axes. Allowance must, of course, be made for the fact that, as the point P moves along the curve, the pair of axes Pt, Pn also move; in particular they rotate in their own plane; when P moves a distance δs along the curve the angle turned through is $\delta\psi$. If $d\psi/ds$ exists at P (and is equal to $1/\rho$) the axes are said to have the "spin" $1/\rho$ at P. For the sake of the kinematic analogy we suppose that P is moving along the curve with uniform unit velocity, so that s may be regarded as representing the time.

To obtain the results of § 3·50 we only require a knowledge of the true rates of increase of the components of a vector which is defined with respect to a system of moving axes. These may be stated in general terms as follows.

Let (x, y) be the vector components (functions of the time t) referred to axes moving with a spin ω, which is positive when it rotates the axis of x towards the axis of y. Let (x', y') be the differential coefficients of (x, y) with respect to t, and $(Dx/Dt, Dy/Dt)$ the true rates of change of the vector components in fixed directions coinciding with the instantaneous directions of the moving axes. Then*

$$(3\text{·}601) \qquad Dx/Dt = x' - y\omega, \quad Dy/Dt = y' + x\omega.$$

Returning to our curve we see that any set of direction cosines defines a

---

* The proof of this classical result is left to the reader.

unit vector whose components may be regarded as functions of the "time" s. The direction cosines of the axes Ox, Oy referred to the moving axes Pt, Pn are respectively

$$(l,\ l'),\ (m,\ m').$$

These directions are fixed in space, and therefore $(l,\ l')$, $(m,\ m')$ are the components of constant vectors. Therefore

$$\frac{Dl}{Ds} = \frac{Dl'}{Ds} = \frac{Dm}{Ds} = \frac{Dm'}{Ds} = 0.$$

Applying 3·601, and replacing ω by $1/\rho$, we find that

$$(3\text{·}602) \qquad \frac{dl}{ds} = \frac{l'}{\rho},\quad \frac{dl'}{ds} = -\frac{l}{\rho},\quad \frac{dm}{ds} = \frac{m'}{\rho},\quad \frac{dm'}{ds} = -\frac{m}{\rho}.$$

But these are equations 3·51 and 3·54, which contain all the results of § 3·50.

§ 3·610. **The kinematics of a rigid body with reference to the properties of twisted curves.** Before leaving the subject of curvature, we propose to sketch shortly the manner in which the properties of a *twisted curve* may be obtained with the help of general geometrical theorems, usually applied to the problems of the kinematics of a rigid body. There is a special set of three lines at right angles through each point of a twisted curve (of which one is the tangent to the curve), referred to which the curve has a specially simple form near the point in question. These three straight lines or triad may be regarded as moving like a rigid body, as their point of intersection moves along the curve, and the nature of the curve is therefore closely connected with the motion of the triad.

The discussion forms a natural extension of the preceding part of the chapter, especially § 3·60. Proofs are, in the main, only outlined, and the reader may, with advantage, construct detailed proofs of any of the theorems enunciated.

It should be borne in mind in what follows that the kinematical language may be regarded simply as illustrative. All terms employed are capable of a purely geometrical or analytical interpretation.

In the first place, any displacement of a rigid body, with one point fixed, may be uniquely represented by a certain rotation about a certain axis through the fixed point, called the axis of rotation*.

Let us now regard the displacement of the body as defined by certain functions of an independent variable (the time t, say) which specify the direction cosines of certain lines fixed in the body referred to axes fixed in space. If, then, the differential coefficients of these functions exist at $t = t_0$, the axis of rotation tends to a limit called the **Instantaneous Axis** as $t \rightarrow t_0$, and the rotation about the axis is asymptotically equal to $\Omega\,(t - t_0)$, where Ω is a definite constant, which may be called the "spin" of the body about the instantaneous axis.

---

* Euler's Theorem. See Routh, *Rigid Dynamics*, Vol. I, Chap. v, § 1; Thomson and Tait, *Natural Philosophy*, Vol. I, Part I, p. 69.

It follows that the motion of the body from t_0 to t can be represented by a rotation $\Omega\,(t-t_0)$ about the instantaneous axis, with an error $o\,\{|\,t-t_0\,|\}*$.

The "spin" obeys the vector laws of composition and resolution. It can be represented as a vector whose direction is the direction of the instantaneous axis and whose magnitude is Ω†. This may be established as follows.

If the body has the spin Ω about an axis whose direction cosines are (l, m, n), a point P in the body, of coordinates (x, y, z), has a velocity whose components are

$$\begin{vmatrix} m & n \\ y & z \end{vmatrix}\Omega, \qquad \begin{vmatrix} n & l \\ z & x \end{vmatrix}\Omega, \qquad \begin{vmatrix} l & m \\ x & y \end{vmatrix}\Omega, \tag{A}$$

along the axes of Ox, Oy, Oz. If we agree to call $(l\Omega, m\Omega, n\Omega)$ the resolved parts of the spin Ω along these axes and write $\omega_1 = l\Omega$, etc., then the components of the velocity of P are

$$-\omega_3 y + \omega_2 z, \quad -\omega_1 z + \omega_3 x, \quad -\omega_2 x + \omega_1 y,$$

which are exactly the same components as are got by considering the body as having *simultaneously* spins ω_1, ω_2, ω_3 about the three axes Ox, Oy, Oz, respectively.

Therefore *spin can be resolved like a vector*, if the direction of this vector is the axis of the spin, and its length proportional to the spin.

In general a body may be said to have simultaneously two spins about any two different axes, if every point P of the body has a velocity which can be represented by the sum of two expressions of type (A). The Ox-component is then

$$-\omega_3 y + \omega_2 z - \omega_3' y + \omega_2' z,$$

or

$$-(\omega_3 + \omega_3')z + (\omega_2 + \omega_2')y.$$

Thus the body is really spinning about an axis and with a spin which is the resultant by the parallelogram law of the components

$$(\omega_1 + \omega_1'), \quad (\omega_2 + \omega_2'), \quad (\omega_3 + \omega_3'),$$

which is identical with the resultant by the same law of (a) $(\omega_1, \omega_2, \omega_3)$ and (b) $(\omega_1', \omega_2', \omega_3')$, *i.e.* of Ω and Ω'.

Therefore *spins may be compounded like vectors*.

The relation between the direction of rotation of the spin Ω and the positive direction of the axis denoted by (l, m, n) is chosen to be such that ω_1, the spin about Ox, is positive when it tends to rotate the axis Oy towards the axis Oz. The same statement is true of the other components when the letters (x, y, z) are cyclically interchanged.

Again suppose that a body, moving about a fixed point O, has an instantaneous axis at $t=t_0$. If two lines fixed in the body have positions OT_1, OT_2 at time t and OT_1', OT_2' at time t_0, and if

$$T_1 \hat{O} T_2' = T_1 \hat{O} T_2 + o\,\{|\,t-t_0\,|\}$$

the instantaneous axis must lie in the plane $T_1 O T_2$‡.

* Routh, *loc. cit.* Art. 217. The necessary extension of this article is obvious.

† Routh, *loc. cit.* Arts. 230–232.

‡ The reader should draw a spherical figure, and apply the preceding properties of the instantaneous axis.

This completes the properties of the spin and instantaneous axis that are required. We want one more property, the analogue for three dimensions of $3 \cdot 601$, concerning the true rates of increase of a vector defined with respect to a system of moving axes.

Let (x, y, z) be the vector components (functions of t) referred to moving axes whose motion may be resolved into spins $(\theta_1, \theta_2, \theta_3)$ about the instantaneous directions of these axes. Let (x', y', z') be the differential coefficients of (x, y, z) with respect to t, and

$$Dx/Dt, \quad Dy/Dt, \quad Dz/Dt$$

the true rates of change of the vector components in fixed directions coinciding with the instantaneous directions of moving axes. Then[*]

$$Dx/Dt = x' - y\theta_3 + z\theta_2,$$
$$Dy/Dt = y' - z\theta_1 + x\theta_3,$$
$$Dz/Dt = z' - x\theta_2 + y\theta_1.$$

This completes the account of the kinematics of a rigid body, so far as required for the discussion of a twisted curve.

§ 3·620. The curvature and torsion of a twisted curve. Frenet's formulae.

The set of fundamental axes associated with a point P of a twisted curve are :—

Axis (1). **The tangent to the curve at** P, drawn in the direction of s increasing. This direction may be arbitrarily chosen for s.

Axis (2). **The principal normal at** P, which is the line through P, in the osculating plane, normal to the tangent at P. The positive direction of the principal normal may be chosen arbitrarily.

Axis (3). **The binormal at** P, which is the normal through P to the osculating plane, drawn in a direction such that the two triads, $P123$ and $Oxyz$, are of the same type, *i.e.* can be placed so that the positive directions $(P1, P2, P3)$, (Ox, Oy, Oz) respectively coincide.

Referred to axes $Oxyz$ let (l_1, m_1, n_1) be the direction cosines of the tangent $P1$, (l_2, m_2, n_2) the direction cosines of the Principal Normal $P2$, (l_3, m_3, n_3) the direction cosines of the Binormal $P3$.

The twisted curve is assumed to be given in the form

$$x = \phi_1(s), \quad y = \phi_2(s), \quad z = \phi_3(s),$$

and $\phi_1', \phi_2', \phi_3'$ are not simultaneously zero.

If $\phi_1''', \phi_2''', \phi_3'''$ exist, and s, for the sake of the kinematic analogy, is regarded as the "time", the triad $P123$, regarded as a rigid body, has an instantaneous axis, which lies in the plane $P13$, *i.e.* normal to $P2$[†].

---

[*] Routh, (*Advanced*) *Rigid Dynamics*, Vol. II, pp. 1–4.

[†] The directions of $P123$ may be defined in terms of $(\phi_1', \phi_2', \phi_3')$ and $(\phi_1'', \phi_2'', \phi_3'')$. Hence the instantaneous axis exists. By the theorem of Ex. 19 β, if $Q1'$ is the direction of the tangent at Q, and $PQ = \delta s$, the angle between $P3$ and $Q1'$ is $\frac{1}{2}\pi + O(\delta s)^2$, and the statement above follows from the last property of the Instantaneous Axis established in § $3 \cdot 610$.

The angular motion of the triad $P123$ may therefore be represented by spins $-1/\tau$, 0, $1/\rho$ about the axes $P1$, $P2$, $P3$. It therefore follows that, if P, Q are two points δs apart, $\delta\psi$ the angle between the tangents at P and Q, and $\delta\epsilon$ the angle between the osculating planes or binormals at P and Q, and if also $(\phi_1''', \phi_2''', \phi_3''')$ exist, then

$$\underset{\delta s \to 0}{\mathrm{Lt}}\ \frac{\delta\psi}{\delta s}=\frac{1}{\rho}, \qquad \underset{\delta s \to 0}{\mathrm{Lt}}\ \frac{\delta\epsilon}{\delta s}=-\frac{1}{\tau}.$$

We define $1/\rho$ and $1/\tau$ to be the **curvature** and **torsion** of the curve at P.

The direction cosines of the axes of reference Ox, Oy, Oz referred to the moving system $P123$ are respectively

$$(l_1, l_2, l_3), \quad (m_1, m_2, m_3), \quad (n_1, n_2, n_3).$$

These directions are fixed in space and therefore, for example, (l_1, l_2, l_3) are the components of a constant vector. Therefore

$$\frac{Dl_1}{Ds}=\frac{Dl_2}{Ds}=\ldots=\frac{Dm_1}{Ds}=\ldots=\frac{Dn_3}{Ds}=0.$$

It follows at once that

$$(3\text{·}621) \qquad 0=\frac{dl_1}{ds}-\frac{l_2}{\rho},$$

$$(3\text{·}622) \qquad 0=\frac{dl_2}{ds}+\frac{l_3}{\tau}+\frac{l_1}{\rho},$$

$$(3\text{·}623) \qquad 0=\frac{dl_3}{ds}-\frac{l_2}{\tau},$$

with similar expressions for the m's and n's. These are **Frenet's Formulae**.

In conclusion we shall establish a few of the more important formulae for ρ and τ which follow from Frenet's formulae.

Since $\Sigma l_1{}^2=1$, $\Sigma l_1 l_2=0$, where Σ refers to summation over the letters $\langle l, m, n \rangle$, with similar formulae for other suffixes, we have at once

$$(3\text{·}624) \qquad \frac{1}{\rho^2}=\Sigma\left(\frac{dl_1}{ds}\right)^2=\underset{x,y,z}{\Sigma}\left(\frac{d^2x}{ds^2}\right)^2,$$

$$(3\text{·}625) \qquad \frac{1}{\tau^2}=\Sigma\left(\frac{dl_3}{ds}\right)^2,$$

$$(3\text{·}626) \qquad \frac{1}{\rho\tau}=\Sigma\frac{dl_1}{ds}\frac{dl_3}{ds}.$$

Again, since (l_3, m_3, n_3) are the direction cosines of the normal to the osculating plane, it follows from Ex. 19, that

$$\frac{l_3}{\begin{vmatrix} m_1 & n_1 \\ m_1' & n_1' \end{vmatrix}}=\frac{m_3}{\begin{vmatrix} n_1 & l_1 \\ n_1' & l_1' \end{vmatrix}}=\frac{n_3}{\begin{vmatrix} l_1 & m_1 \\ l_1' & m_1' \end{vmatrix}}=\frac{\Sigma l_3{}^2}{\begin{vmatrix} l_3 & m_3 & n_3 \\ l_1 & m_1 & n_1 \\ l_1' & m_1' & n_1' \end{vmatrix}}=\frac{1}{\dfrac{1}{\rho}\begin{vmatrix} l_1 & m_1 & n_1 \\ l_2 & m_2 & n_2 \\ l_3 & m_3 & n_3 \end{vmatrix}}=\rho,$$

for the triad $P123$ or (l_1, m_1, n_1), (l_2, m_2, n_2), (l_3, m_3, n_3) is of the same type as $Oxyz$. Dashes denote differentiation with respect to s. Therefore

$$\frac{1}{\rho\tau} = \Sigma l_1' l_3' = \Sigma l_1' \rho' \begin{vmatrix} m_1 & n_1 \\ m_1' & n_1' \end{vmatrix} + \rho \Sigma l_1' \begin{vmatrix} m_1 & n_1 \\ m_1'' & n_1'' \end{vmatrix} = \rho \begin{vmatrix} l_1' & m_1' & n_1' \\ l_1 & m_1 & n_1 \\ l_1'' & m_1'' & n_1'' \end{vmatrix}.$$

Hence

$$(3\text{·}627) \qquad \frac{1}{\rho^2 \tau} = - \begin{vmatrix} l_1 & m_1 & n_1 \\ l_1' & m_1' & n_1' \\ l_1'' & m_1'' & n_1'' \end{vmatrix} = - \begin{vmatrix} x' & y' & z' \\ x'' & y'' & z'' \\ x''' & y''' & z''' \end{vmatrix}.$$

§ 3·70. **Evolutes and involutes**[*]. From here onwards we shall need to assume more than the existence of ρ, and shall therefore assume the existence and continuity of all differentials or derivates that are mentioned. It is easy to see that, for the curve $y = f(x)$, the continuity of $d\rho/ds$ is equivalent to the continuity of $f'''(x)$, and so on. Such details are here of no great interest or importance.

DEFINITIONS. **The evolute** *of a given curve is the locus of its centre of curvature. Any curve which has a given curve for evolute is called* **an involute** *of the given curve.*

Thus a given curve is an involute of its evolute, but we shall see it is only one of many.

We shall make a rule of using capital letters for points, etc. belonging to the evolute, X, Y, P, S, Ψ, L, M, L', M' (if required) having the same meanings for the evolute that x, y, ρ, s, ψ, l, m, l', m' have for the original curve, and all letters referring to corresponding points. Now we have

$$X = x + l'\rho, \quad Y = y + m'\rho.$$

Therefore

$$(3\text{·}701) \qquad \begin{cases} dX = dx + l'd\rho + \rho dl' = l'd\rho, \\ dY = dy + m'd\rho + \rho dm' = m'd\rho, \end{cases}$$

so that

$$(3\text{·}71) \qquad (dS)^2 = (d\rho)^2.$$

So long as $d\rho$ does not change sign, we can choose a suitable direction in which to measure S on the evolute, and have

$$(3\text{·}711) \qquad dS = d\rho \,[\dagger].$$

---

[*] Picard, *Traité d'Analyse*, 2nd Ed., Vol. I, pp. 350 sqq. We shall in future refer to this book as Picard.

[†] This is of course an exact relation.

Now we have as usual

$$(3\text{·}712) \qquad dX = L\,dS, \quad dY = M\,dS,$$

and as a consequence of 3·701 and 3·712 we see that $L = l'$, $M = m'$. We have thus established the following properties.

THEOREM 3·72. *The evolute touches each normal at the centre of curvature.*

THEOREM 3·73. *The arc of the evolute, corresponding to an arc of the original curve for which ρ constantly increases or decreases, is equal to the difference of the radii of curvature touching its extremities.*

The last theorem follows at once by integrating 3·711; for we thus get $S = \rho + c$, where c is a constant.

It follows at once from 3·711 that every plane curve whose curvature is constant is a circle, for, since $(dS)^2 = 0$, the centre of curvature at one point is the centre of curvature at every point of the curve. Therefore every point of the curve lies on a fixed circle.

A formula for P is easily found. For we have

$$\mathrm{P} = (dS/d\Psi) = (d\rho/d\Psi),$$

and it is evident that $|\,d\Psi\,| = |\,d\psi\,|$. Hence, without regard to sign,

$$(3\text{·}74) \qquad \mathrm{P} = \frac{d\rho}{ds}\frac{ds}{d\psi} = \rho\,\frac{d\rho}{ds}.$$

The converse of Theorem 3·72 is also true.

THEOREM 3·75. *A curve which at every point touches a normal to a given curve Γ is the evolute of Γ (or part of it).*

This is, properly speaking, a particular case of Theorem 5·32 on envelopes, but a direct proof is easy. For if a normal to Γ touches the curve at C, then C is the limit of the point of intersection of neighbouring normals, and therefore a centre of curvature of Γ. Hence all points on the curve are centres of curvature of Γ, which proves the theorem.

§ 3·80. THEOREM 3·81. *To a given curve there belong infinitely many involutes.*

Let $P(x, y)$ be a point on a given curve, and $Q(\xi, \eta)$ a point on the tangent at P such that $PQ = -(s + a)$, where a is any constant. Then

$$\xi = x - (s + a)\,l, \quad d\xi = -(s + a)\,dl,$$

$$\eta = y - (s + a)\,m, \quad d\eta = -(s + a)\,dm,$$

and therefore $\qquad d\xi : d\eta = l' : m',$

so that the tangent at Q to the locus of Q is parallel to the normal at P; *i.e.* QP is the normal to the locus of Q at Q. Hence the given curve

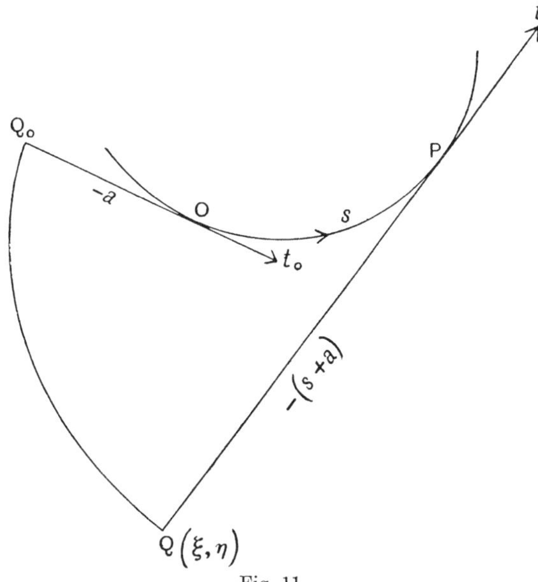

Fig. 11.

touches all the normals to the locus of Q, and each of its points is a point of contact with some normal. Therefore, by Theorem 3·75, the given curve is the evolute of the locus of Q, which is therefore an involute. It is clear that we get a new involute for each value of a.

The following mechanical description of the involute is of interest. Suppose that a thread, inextensible, perfectly flexible, and without thickness, is wound tightly round an arc of the curve, leaves the curve along the tangent at O, and ends on any chosen involute at Q_0. If the thread be then unwound off the curve, being always kept tight, the end will describe the chosen involute. For when the thread leaves the curve at P, and the end is at Q along the tangent at P, we have

$$QP - Q_0O = \text{arc } OP,$$

and therefore Q is still on the involute through Q_0. Hence the end describes the chosen involute.

EXAMPLES II

(1) Prove that, with due regard to sign,

$$\rho = r \, dr/dp$$

in all cases, p and r being the tangential-polar coordinates of the curve.

[Lamb, *Infinitesimal Calculus*, 2nd Ed., pp. 401–402.]

(2) By drawing the normals at points PQR such that arc $PQ=$ arc $QR=\delta s$, prove that the evolute touches the normals and that

$$dS=d\rho.$$

[Let $X_1 X_2$ be the mid-points of the arcs PQ, QR, and $C_1 C_2$ their centres of curvature; then C_1 is represented with an error $O\,(\delta s)^2$ by Γ_1, the intersection of normals at P and Q, and C_2 similarly by Γ_2, the intersection of normals at Q and R. Hence $\Gamma_1 \Gamma_2 \sim C_1 C_2$, and $\Gamma_1 \Gamma_2$ tends to parallelism with the tangent to the evolute at C_1. Hence the theorem.]

(3) Taking the tangent and normal to a curve as the axes of x and y, and measuring s from the origin in the direction of x increasing, prove that

$$x=s-\frac{s^3}{3!}\left(\frac{d\psi}{ds}\right)_0^2+O\,(s^4),$$

$$y=\frac{s^2}{2!}\left(\frac{d\psi}{ds}\right)_0+\frac{s^3}{3!}\left(\frac{d^2\psi}{ds^2}\right)_0+O\,(s^4).$$

In particular if

$$\left(\frac{d\psi}{ds}\right)_0=\left(\frac{d^2\psi}{ds^2}\right)_0=\ldots=\left(\frac{d^{r-1}\psi}{ds^{r-1}}\right)_0=0,$$

$(d^r\psi/ds^r)_0 \neq 0$, and $d^{r+1}\psi/ds^{r+1}$ continuous, then

$$x=s-\frac{s^{2r+1}}{2r\,(4r^2-1)\,\{(r-1)\,!\}^2}\left(\frac{d^r\psi}{ds^r}\right)_0^2+O\,(s^{2r+2}),$$

$$y=\frac{s^{r+1}}{(r+1)\,!}\left(\frac{d^r\psi}{ds^r}\right)_0+O\,(s^{r+2}).$$

[We have $x=\displaystyle\int_0^s \cos\psi\,ds$, $y=\displaystyle\int_0^s \sin\psi\,ds$. Expand these expressions by Taylor's theorem, remembering that $\psi_0=0$.]

(4) If PT, QT are the tangents at P and Q, meeting in T, and the conditions of the latter part of (3) hold at P, then $PT/QT \to r$ as $Q \to P$.

[Use the result of Example (3) above.]

(5) If PT, QT are the tangents at P and Q, meeting in T, the circle PQT has, as its limit as $Q \to P$, a circle of radius $\tfrac{1}{2}\rho$ whose centre lies on the normal at P.

[If the normals at P and Q meet at K, $PQTK$ are concyclic.]

(6) If T_1, T_2, T_3 are the three points of intersection of the tangents at P, Q_1, Q_2, the circle $T_1 T_2 T_3$ has, as its limit as Q_1, $Q_2 \to P$, a circle of radius $\tfrac{1}{4}\rho$ whose centre lies on the normal at P.

[Use Theorem 2·624.]

CHAPTER IV

THE THEORY OF CONTACT*

§ 4·10. **The distance from a curve of a point near it.** If two curves have a common point P, it is a matter of some interest in itself, and of vital importance for further developments, to investigate how close the two curves lie to one another in the neighbourhood of P.

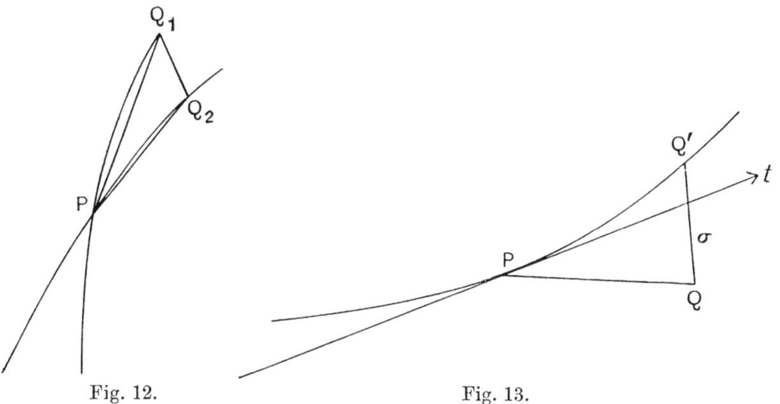

Fig. 12. Fig. 13.

More precisely, it is required to determine the order of $Q_1 Q_2$ (Fig. 12) when $Q_1 \to P$ and $Q_2 \to P$, *i.e.* when $Q_1 P$ and $Q_2 P$ are small lengths of the first order of smallness. With this object in view we proceed as follows.

We take the curve in the form

$$f(x, y) = 0,$$

and suppose that $P(a, b)$ is an *ordinary* point of the curve, *i.e.* that f_a' and f_b' are not both zero. Let Q be any point (x, y) near P, *not* on the curve, and Q' any point (x', y') near P on the curve, so that

$$f(a, b) = f(x', y') = 0.$$

* d.l.V.P., Vol. ii, p. 396 (on whose exposition our treatment is based); Picard, Vol. i, p. 342; Goursat, Vol. i, p. 530; Jordan, Vol. i, p. 417 (a treatment which includes complex points and curves).

Let the length and direction cosines of QQ' be σ and $(l,\ m)$. We proceed to determine an asymptotic formula for σ valid as $Q \to P$. We assume that $f(x, y)$ has as many continuous differential coefficients as may be mentioned. At present our requirements are second order differential coefficients in the neighbourhood of P. We have

$$x' = x + l\sigma, \quad y' = y + m\sigma,$$
$$0 = f(x + l\sigma,\ y + m\sigma),$$
$$= f(x,\ y) + \sigma\,(lf_\xi' + mf_\eta')^*,$$

where ξ is some number between x and x' and η some number between y and y'. Now when $Q \to P$ in any manner, l and m remaining constant and equal to l_0 and m_0, or satisfying the relations $l \to l_0$, $m \to m_0$, we have

$$f(x,\ y) \to f(a,\ b) = 0,$$

(4·101) $$lf_\xi' + mf_\eta' \to l_0 f_a' + m_0 f_b'.$$

If therefore $$l_0 f_a' + m_0 f_b' \neq 0,$$

i.e. if QQ' is parallel to, or tends to parallelism with, a fixed line L *not parallel to the tangent at P†*, then

(4·102) $$\sigma \to 0,$$

as $Q \to P$.

Again we have

$$0 = f(x,\ y) + \sigma\,(lf_x' + mf_y') + O\,(\sigma^2)\ddagger,$$
$$= f(x,\ y) + \sigma\,\{(l_0 f_a' + m_0 f_b') + o\,(1)\},$$

by 4·101 and 4·102, and, since

$$l_0 f_a' + m_0 f_b' \neq 0,$$

we have $$\sigma \sim \frac{-f(x,\ y)}{l_0 f_a' + m_0 f_b'},$$

as $Q \to P$. We can therefore enunciate the following theorem.

THEOREM 4·11. *If P is an ordinary point on the curve $f(x,\ y) = 0$, and σ is the distance from a neighbouring point $Q\,(x,\ y)$ to the curve, measured parallel§ to a straight line not parallel to the tangent at P, then*

$$\sigma \sim A f(x,\ y),$$

where $A \neq 0\|$, as $Q \to P$.

---

* By Taylor's Theorem, for $n = 1$.

† This condition is sufficient (but not necessary) to ensure that $\sigma \to 0$.

‡ By Taylor's Theorem, for $n = 2$.

§ Or, '*in a direction that tends to parallelism with*'.

‖ Or, in the notation of *Orders of Infinity*, p. 2,

$$\sigma \underset{\sim}{\asymp} f(x,\ y).$$

The following alternative form may sometimes be more convenient.

THEOREM 4·12. *If σ is the shortest distance from Q to the curve, the other conditions of Theorem 4·11 remaining unaltered, then*

$$\sigma \sim -f(x, y)/(f_a'^2 + f_b'^2)^{\frac{1}{2}},$$

as $Q \to P$.

If QQ' is the shortest distance from Q to the curve, it is easily proved that QQ' is normal to the curve at Q'. As $Q \to P$, $\sigma \to 0$, and therefore $Q' \to P$, so that the direction parallel to which QQ' is measured tends to parallelism with the normal at P. Hence

$$l \to f_a'/(f_a'^2 + f_b'^2)^{\frac{1}{2}}, \quad m \to f_b'(f_a'^2 + f_b'^2)^{\frac{1}{2}},$$

and the theorem follows at once.

§ 4·20. **Definition of contact of order** n. We now make the following definition.

DEFINITION*. *Two curves that have an ordinary point P in common are said to have* **contact of order** n *at* P, *if the distance† of a point Q of one curve from the other is of the $(n+1)$th order of smallness compared to QP.*

They may be said to have **contact at least of order** n *if the distance is at least of the $(n+1)$th order of smallness compared to QP.*

Another way of stating the same thing is, of course, to say that if σ is the distance in question, then

$$\sigma \asymp (QP)^{n+1},$$

or in the second case $\qquad \sigma \preccurlyeq (QP)^{n+1}.$

Suppose that the curve on which Q is not taken has the equation $f(x, y) = 0$. Then the results of the last section may be expressed as follows.

THEOREM 4·21. *The necessary and sufficient condition that the two curves should have contact of order n at the point P is that, when the coordinates of Q are substituted for x and y in $f(x, y)$, the expression $f(x, y)$ should satisfy the relation*

$$(4·211) \qquad\qquad f(x, y) \asymp (PQ)^{n+1},$$

as $Q \to P$ *along its curve.*

---

* It should be observed that the definition as it stands needs justification, for it defines a property symmetrical with respect to the two curves in terms of an unsymmetrical property. But it will appear that the conditions of contact are symmetrical, so that Q may be taken on either curve, which affords the necessary justification.

† Distance may be taken to mean either (1) shortest distance, or (2) distance measured parallel to a line not parallel to the tangent of the other curve at P.

The following special cases are of the greatest importance.

I. Suppose that Q lies on the curve
$$x = \phi_1(t), \quad y = \phi_2(t),$$
that the other curve is
$$f(x, y) = 0,$$
that P is the point of parameter t_0 on Q's curve, and that P is an ordinary point on both curves. Then, if Q is the point of parameter t,
$$PQ \gtreqqless |t - t_0|,$$
as $Q \to P$ or $t \to t_0$. Following the general rule, it is necessary and sufficient for contact of order n that
$$f\{\phi_1(t), \phi_2(t)\} \gtreqqless |t - t_0|^{n+1}.$$
Writing
$$f\{\phi_1(t), \phi_2(t)\} \equiv \Phi(t),$$
and applying Taylor's Theorem, it appears that if $\Phi(t)$ has $n + 1$ continuous differential coefficients at $t = t_0$, *the necessary and sufficient conditions for contact of order n are that*

$$(4\text{·}212) \qquad \Phi(t_0) = \Phi'(t_0) = \ldots = \Phi^{(n)}(t_0) = 0, \quad \Phi^{(n+1)}(t_0) \neq 0.$$

II. In particular suppose that the two curves are
$$y = f_1(x), \quad y = f_2(x).$$
If we take the latter for Q's curve, and $x = t$ as the parameter, then
$$\Phi(t) \equiv f_2(x) - f_1(x);$$
and, if the parameter of P is x_0, *the necessary and sufficient conditions are*

$$(4\text{·}213) \qquad f_2(x_0) = f_1(x_0), \quad f_2'(x_0) = f_1'(x_0), \quad \ldots,$$
$$f_2^{(n)}(x_0) = f_1^{(n)}(x_0), \quad f_2^{(n+1)}(x_0) \neq f_1^{(n+1)}(x_0).$$

It should be noticed that these conditions are symmetrical with respect to the two curves. Moreover this form of equation is perfectly general unless the tangent to the curves at P is parallel to the axis of y. But in this case the equations may be given in the form
$$x = f_1(y), \quad x = f_2(y),$$
leading, as before, to symmetrical conditions. It is therefore a matter of indifference on which curve Q is taken, and the symmetrical form of our original definition is justified. We note in passing the following theorems.

THEOREM 4·22. *Two curves which have contact of order n (i.e. of order n and of no higher order) at a point P cross or do not cross at P according as n is even or odd.*

THEOREM 4·23. *Two curves which have contact of order n with a third curve, have contact* **at least** *of order n with each other.*

The preceding treatment does not apply directly to the case of two curves whose equations are both given in parametric form. As this case seldom occurs in practice, and the preceding discussion is theoretically complete, we shall content ourselves here with a reference to Picard, Vol. I, pp. 342 sqq. The case of the two curves $F_1(x, y) = 0$, $F_2(x, y) = 0$ is considered by Jordan, *loc. cit.*, p. 420.

§ 4·30. **Osculating curves.** Given a curve and an ordinary point P thereon, and any family of curves depending on $n + 1$ parameters, say

$$(4·301) \qquad f(x, y, a_1, a_2, \ldots, a_{n+1}) = 0,$$

suppose that a member T of this family is chosen by determining the parameters so that T has contact of the highest possible order for the family with the given curve at the point P. Then T *is said to have* **osculating contact** *with the given curve at the point* P: the two curves are also said *to osculate* or *to be osculating curves.*

Suppose that the given curve is

$$x = \phi_1(t), \quad y = \phi_2(t),$$

and that P is the point t_0. The conditions for contact of order n at the point P are by 4·212

$$\Phi(t_0) = \Phi'(t_0) = \ldots = \Phi^{(n)}(t_0) = 0, \quad \Phi^{(n+1)}(t_0) \neq 0,$$

where $\qquad \Phi(t) \equiv f\{\phi_1(t), \phi_2(t), a_1, a_2, \ldots, a_{n+1}\}.$

In general these $n + 1$ equations will be just sufficient to determine the $n + 1$ parameters in the equation of the family (not necessarily uniquely); and *in general* it will not be the case that, when the a's have been so determined, $\Phi^{(n+1)}(t_0) = 0$. Therefore *in general the curve* (*or curves*) *of the family*

$$(4·301) \qquad f(x, y, a_1, a_2, \ldots, a_{n+1}) = 0$$

that has (*have*) *osculating contact with a given curve at the point* P, *has* (*have*) *contact of order n at that point.*

The parameters of an osculating curve T are determined by $n + 1$ equations which written in full are

$$(4·302) \quad \begin{cases} \Phi(t_0, a_1, a_2, \ldots, a_{n+1}) = 0, \\ \Phi'(t_0, a_1, a_2, \ldots, a_{n+1}) = 0, \\ \cdots\cdots\cdots\cdots\cdots\cdots\cdots\cdots \\ \Phi^{(n)}(t_0, a_1, a_2, \ldots, a_{n+1}) = 0. \end{cases}$$

We shall denote a possible set of real values of the parameters, satisfying these equations, by a_1, a_2, ..., a_{n+1}, and the corresponding osculating curve by T. Suppose now that K is a curve of the family which passes through P*, and n neighbouring points on the given curve whose parameters are t_0, t_1, t_2, ..., t_n. The parameters (a's) must satisfy the equations

(4·303) $\Phi(t_r, a_1, a_2, ..., a_{n+1}) = 0$, $(r = 0, 1, 2, ..., n)$.

Without loss of generality we may suppose that $t_0 < t_1 < ... < t_n$. Then, by Rolle's Theorem, there are at least n distinct values of t, between t_0 and t_n, for which $\Phi'(t, a_1, a_2, ..., a_{n+1}) = 0$; by a second application of Rolle's Theorem there must therefore be at least $n-1$ distinct values of t, between t_0 and t_n, for which $\Phi''(t_1, a_1, a_2, ..., a_{n+1}) = 0$, and so finally at least one value of t, between t_0 and t_n, such that

$$\Phi^{(n)}(t, a_1, a_2, ..., a_{n+1}) = 0.$$

The system of equations 4·303 may therefore be replaced by the equivalent system

(4·304)
$$\begin{cases} \Phi(t_0, a_1, a_2, ..., a_{n+1}) = 0, \\ \Phi'\{(t_0 + \mu_1), a_1, a_2, ..., a_{n+1}\} = 0, \\ \cdots\cdots\cdots\cdots\cdots\cdots\cdots\cdots\cdots \\ \Phi^{(n)}\{(t_0 + \mu_n), a_1, a_2, ..., a_{n+1}\} = 0, \end{cases}$$

where μ_1, μ_2, ..., $\mu_n \to 0$ as t_1, t_2, ..., $t_n \to t_0$, i.e. as

$$Q_1, Q_2, ..., Q_n \to P.$$

The equations 4·304 may be regarded as a system of $n+1$ equations determining the a's as functions of the μ's. If the Jacobian† of this system does not vanish when

$$\mu_1 = \mu_2 = ... = \mu_n = 0, \quad a_1 = a_1, ..., a_{n+1} = a_{n+1},$$

that is *in general*, the system 4·304 determines a unique set of functions θ_1, θ_2, ..., θ_{n+1}, such that

$$a_r = a_r + \theta_r, \quad (r = 1, 2, ..., n+1),$$

and θ_r is a function of μ_1, μ_2, ..., μ_n, real and continuous when

$$|\mu_r| < \delta, \quad (r = 1, 2, ..., n),$$

and such that $\theta_r \to 0$, $(r = 1, 2, ..., n)$,

when $\mu_1, \mu_2, ..., \mu_n \to 0$.

---

* It is convenient to take this form of hypothesis, but not necessary. K may be taken to be a curve through $Q_1, Q_2, ..., Q_{n+1}$, $n+1$ points near P, and $Q_1, Q_2, ..., Q_{n+1}$ made to tend to P.

† Goursat, Vol. I, Chap. III, in particular, pp. 96–97.

It follows that, if Q_1, Q_2, ..., Q_n be sufficiently near P, a unique* curve K of the family can be drawn through P, Q_1, Q_2, ..., Q_n, which is such that, when Q_1, O_2, ..., $Q_n \to P$, K has the limit T.

In particular if the parameters appear in $f(x, y, a_1, ..., a_{n+1})$ in such a way that a_1, a_2, ..., a_{n+1} are determined uniquely by the conditions for contact of order n, as is usually the case in practice, the curve K will be genuinely unique and tend to the unique limit T.

With the reservations that the foregoing analysis has disclosed, we can enunciate the following theorem.

THEOREM 4·31. *In general, the member of the family 4·301, which has osculating contact with a given curve at P, is a limit of curves of the family which pass through P and n neighbouring points on the given curve.*

This theorem enables us to attach a meaning to, and renders permissible, the use of such statements as " Two curves cut at P in n coincident points", statements which are formulated for the sake of generality, *e.g.* in order to enable us to say that any two conics have four common points real or complex. Their permissibility being established, such statements are frequently useful and illuminating.

§ 4·40. **Examples of osculating curves.** We have defined a tangent at P as the limit of the chord PQ when $Q \to P$. Since any straight line can be put in the form $ax + \beta y - 1 = 0$, the osculating straight line at the point P is by the last theorem the tangent.

We can of course arrive at this fact directly as follows. Suppose the given curve is

$$x = \phi_1(t), \quad y = \phi_2(t):$$

then $$\Phi(t) \equiv a\phi_1(t) + \beta\phi_2(t) - 1 ;$$

and the conditions for first order contact at t_0 are

$$a\phi_1(t_0) + \beta\phi_2(t_0) - 1 = 0, \quad a\phi_1'(t_0) + \beta\phi_2'(t_0) = 0,$$

so that the osculating straight line is

$$\begin{vmatrix} x & y & 1 \\ \phi_1(t_0) & \phi_2(t_0) & 1 \\ \phi_1'(t_0) & \phi_2'(t_0) & 0 \end{vmatrix} = 0,$$

which is the tangent. It should be remembered that, t_0 being an ordinary point, $\phi_1'(t_0)$ and $\phi_2'(t_0)$ are not both zero.

---

* It must not be supposed that K is necessarily the only curve of the system through P, $Q_1, ... Q_n$. K is the only curve of this nature whose parameters are "nearly equal to the a's".

We should also observe that *in general* the curve does not cross its tangent. If however it happens that the contact is of the second order at t_0, *i.e.* if also

$$a\phi_1''(t_0) + \beta\phi_2''(t_0) = 0,$$

or

$$\begin{vmatrix} \phi_1'(t_0) & \phi_2'(t_0) \\ \phi_1''(t_0) & \phi_2''(t_0) \end{vmatrix} = 0,$$

so that t_0 is a point of inflexion, the curve will, in general, cross its tangent.

In §§ 3·10, 3·20 we defined the circle of curvature at a point $P(t_0)$ and proved that, if $\phi_1''(t)$ and $\phi_2''(t)$ are continuous at t_0, the circle of curvature is the limit of a circle passing through P and any neighbouring points Q_1 and Q_2, when Q_1, $Q_2 \to P$. The equation of any circle can be put in the form

$$(x-a)^2 + (y-b)^2 - R^2 = 0,$$

and a, b, R^2 are determined uniquely by the conditions of second order contact, except at a point of inflexion*.

It follows from the last theorem (as for a tangent) that the circle of curvature and the osculating circle are identical.

Or directly, for the curve $y = f(x)$, supposing that $f''(x) \neq 0$, we must have

$$(4\cdot41) \qquad \begin{cases} (x-a)^2 + \{f(x) - b\}^2 - R^2 = 0, \\ x - a + f'(x)\{f(x) - b\} = 0, \\ 1 + f'^2(x) + f''(x)\{f(x) - b\} = 0, \end{cases}$$

so that

$$f(x) - b = -\frac{1 + f'^2(x)}{f''(x)},$$

$$x - a = \frac{f'(x)\{1 + f'^2(x)\}}{f''(x)},$$

$$R^2 = \{1 + f'^2(x)\}^3 / f''^2(x);$$

which agree with the equations of § 3·10. We notice that the circle of curvature has in general second order of contact, and therefore crosses the curve except at points at which, besides 4·41, the equation

$$3f''f' + f'''(f - b) = 0$$

is satisfied. These are the points at which

$$(4\cdot42) \qquad 3f''^2 f' - (1 + f'^2)f''' = 0;$$

and it is easily verified that 4·42 is the same as $dR/ds = 0$.

A direct corollary of Theorem 4·23 may be noticed, namely that two curves which have second order contact at any point have the same osculating circle at that point, and so the same radius of curvature.

§ 4·430. **Extension of the theory of osculating curves.** In § 4·30 we defined those members of a given family of curves which have osculating contact with a given curve at P, and showed that, in general, any such

---

* At a point of inflexion the tangent has second order contact and the curvature is zero. This may be conveniently expressed by saying that "the radius of curvature is infinite and that the circle of curvature is the tangent".

osculating curve is the limit of members of the family which have $n+1$ points of intersection with the given curve, when these $n+1$ points of intersection tend to P. This property may be extended as follows. Let

$$(4\text{·}431) \qquad g\,(x,\,y,\,a_1,\,a_2,\,...,\,a_{n-\lambda})=0$$

be a family of curves, depending on $n-\lambda$ parameters, all of which curves have contact at least of order λ with a given curve $x=\phi_1\,(t)$, $y=\phi_2\,(t)$, at the point $P\,(t_0)$. Then, as before, we say that the osculating curve* of the family is that member of the family which has contact of the highest possible order with the given curve at P. This osculating curve* will, in general, have contact of order n and, if

$$\Phi\,(t)=g\,\{\phi_1\,(t),\,\phi_2\,(t),\,a_1,\,a_2,\,...,\,a_{n-\lambda}\},$$

the conditions for contact of order n which determine the a's are

$$(4\text{·}432) \qquad \Phi^{(\lambda+r)}\,(t_0)=0, \quad (r=1,\,2,\,...,\,n-\lambda).$$

It will be observed that

$$\Phi\,(t_0)=\Phi^{(r)}\,(t_0)=0, \quad (r=1,\,2,\,...,\,\lambda),$$

for all values of the parameters, since all members of the family have contact at least of order λ. We can now argue exactly as before in § 4·30 and arrive at the following theorem.

THEOREM 4·433. *In general, the member of the family* 4·431, *which has osculating contact with a given curve at P, has contact of order n with the given curve, and is the limit of curves of the family which [have contact of order λ at P and] pass through $n-\lambda$ points on the given curve in the neighbourhood of P.*

As an example, the foregoing theorem may be used to identify the Newtonian circle of curvature with the osculating circle. The Newtonian circle of curvature (§ 3·40) is the limit as $Q \to P$ of the circle touching a given curve at P and cutting it at Q. The foregoing discussion shows that this limit is a circle with second order contact with the given curve at P, that is, as such a circle is unique, the osculating circle.

§ 4·50. **Similar problems in three dimensions.** The treatment of the distance of a point Q from a surface

$$f\,(x,\,y,\,z)=0,$$

when Q is near an ordinary point P of the surface, is substantially the same as the treatment for plane curves. We obtain the following theorem.

THEOREM 4·51. *If $P\,(a,\,b,\,c)$ is an ordinary point† on the surface*

$$f\,(x,\,y,\,z)=0,$$

and σ is the distance from a neighbouring point $Q\,(x,\,y,\,z)$ to the surface,

---

* Or, curves.

† That is, a point such that $f_a{}',\,f_b{}',\,f_c{}'$ are not simultaneously zero.

measured parallel to a straight line not parallel to the tangent plane at P, then*
$$\sigma \sim A f(x, y, z),$$
where $A \neq 0$, as $Q \to P$.

Under the same conditions, if σ is the shortest distance,
$$\sigma \sim -f(x, y, z)/(f_a'^2 + f_b'^2 + f_c'^2)^{\frac{1}{2}},$$
as $Q \to P$.

On the other hand, a slight complication is introduced into the treatment of a twisted curve
$$f(x, y, z) = 0, \quad g(x, y, z) = 0$$
by the fact that there are *two* equations. Defining P, Q, Q', and (l, m, n) as in § 4·10, P and Q' lie on both the surfaces $f = 0$ and $g = 0$, and Q lies off at least one surface, say $f = 0$. It is moreover assumed that QQ' is not parallel to, and does not tend to parallelism with, the tangent to the twisted curve at P. If P is an ordinary point, and (u, v, w) are the direction cosines of the tangent to the curve at P, (u, v, w) satisfy and are determined by the equations

$(4·511) \qquad uf_a' + vf_b' + wf_c' = 0, \quad ug_a' + vg_b' + wg_c' = 0\dagger.$

Since by hypothesis (l, m, n) or the limits of (l, m, n) do not satisfy the relations
$$u/l = v/m = w/n,$$
at least one of the expressions
$$lf_a' + mf_b' + nf_c', \quad lg_a' + mg_b' + ng_c'$$
is not zero, and does not tend to zero ; in particular if Q lies on $g = 0$, but not on $f = 0$,
$$lf_a' + mf_b' + nf_c' \neq 0\dagger.$$
We then prove that $\sigma \to 0$, and then that
$$f(x, y, z) + \sigma\{lf_a' + mf_b' + nf_c' + o(1)\} = 0,$$
$$g(x, y, z) + \sigma\{lg_a' + mg_b' + ng_c' + o(1)\} = 0.$$

* Or, in a direction that tends to parallelism with.

† Since $f = 0$ and $g = 0$ do not touch at P. Points at which $f = 0$ and $g = 0$ touch cannot be regarded as ordinary points on the curve of intersection. The equations $f = 0$ and $g = 0$ may be regarded as a pair of simultaneous equations determining say x and y as functions of z. Such a determination requires that the Jacobian
$$\partial(f, g)/\partial(x, y)$$
should not vanish. At a point of contact of $f = 0$ and $g = 0$ all the three Jacobians
$$\partial(f, g)/\partial(x, y), \quad \partial(f, g)/\partial(y, z), \quad \partial(f, g)/\partial(z, x)$$
vanish and no pair of coordinates can be determined by the usual theorems as functions of the third.

We thus have two expressions for σ of which one at least gives an intelligible result.

If for example, $lg_a' + mg_b' + ng_c'$ is equal to or tends to 0, $g\,(x, y, z)$ will be of a higher order of smallness than $f\,(x, y, z)$ [*i.e.* $g/f \rightarrow 0$], and then $\sigma \sim Af\,(x, y, z)$ is correct and $\sigma \sim Ag\,(x, y, z)$ is false. We sum up in the following theorem.

THEOREM 4·52. *With the notation of Theorem* 4·51, *the distance* σ *from an ordinary point of the curve*

$$f\,(x, y, z) = 0, \quad g\,(x, y, z) = 0$$

satisfies whichever of the relations

$$\sigma \sim Af\,(x, y, z), \quad \sigma \sim A'g\,(x, y, z)$$

gives a greater value to σ *(i.e. the lower order of smallness).*

We can now extend our definition of contact of order n to any pair of curves plane or twisted or to a curve and a surface. *They are said to have contact of order n at a common point P if the shortest distance from a neighbouring point Q on one curve to the other curve or to the surface is of the $(n + 1)$th order of smallness compared to PQ.*

§ 4·60. **Contact of a curve and a surface. Osculating surfaces.** After § 4·50 the following statements offer no difficulty.

THEOREM 4·61. *The necessary and sufficient condition that a curve and a surface $f\,(x, y, z) = 0$ should have contact of order n at a common point P is that, when the coordinates of a neighbouring point Q on the curve are substituted for (x, y, z) in $f\,(x, y, z)$, $f\,(x, y, z)$ should be of the $(n + 1)$th order of smallness compared to PQ.*

If the curve is given by

$$x = \phi_1\,(t), \quad y = \phi_2\,(t), \quad z = \phi_3\,(t),$$

and

$$\Phi\,(t) \equiv f\{\phi_1\,(t), \quad \phi_2\,(t), \quad \phi_3\,(t)\},$$

and P is the point t_0, the above condition reduces to

$$(4·611) \qquad \Phi\,(t_0) = \Phi'\,(t_0) = \ldots = \Phi^{(n)}\,(t_0) = 0,$$

together with the inequality $\Phi^{(n+1)}\,(t_0) \neq 0$.

By taking $z = f\,(x, y)$ for the surface and $z = f_2\,(x)$, $y = f_1\,(x)$ for the curve, we can prove that *if a curve has contact of order n (and so of no higher order) with a surface, it crosses or does not cross the surface according as n is even or odd.*

Given a surface depending on $n + 1$ parameters, we can define the osculating surface to a given curve at a given point, and prove that it *is in general the limit of surfaces of its type which intersect the given curve at P and n neighbouring points.*

§ 4·70. **Contact of two twisted curves. Osculating curves.**
Suppose one of the curves is given in the form

$$f(x, y, z) = 0, \quad g(x, y, z) = 0,$$

and consider an ordinary point P. The reader can easily prove the following theorem.

THEOREM 4·71. *The necessary and sufficient conditions that two twisted curves, one of which is given by $f(x, y, z) = 0$, $g(x, y, z) = 0$, should have contact at least of order n at $P*$ is that, when the coordinates of a point Q on the other curve are substituted for (x, y, z) in $f(x, y, z)$ and $g(x, y, z)$, both these expressions should be of the $(n + 1)$th order of smallness at least compared to PQ.*

If the contact is of order n, one at least of $f(x, y, z)$ and $g(x, y, z)$ must be of the $(n + 1)$th order of smallness exactly.

If the other curve is given in the form

$$x = \phi_1(t), \quad y = \phi_2(t), \quad z = \phi_3(t),$$

and

$$\Phi(t) \equiv f\{\phi_1(t), \phi_2(t), \phi_3(t)\},$$

$$\Psi(t) \equiv g\{\phi_1(t), \phi_2(t), \phi_3(t)\},$$

then these conditions reduce to

$$(4·711) \quad \begin{cases} \Phi(t_0) = \Phi'(t_0) = \ldots = \Phi^{(n)}(t_0) = 0, \\ \Psi(t_0) = \Psi'(t_0) = \ldots = \Psi^{(n)}(t_0) = 0, \end{cases}$$

while one at least of

$$\Phi^{(n+1)}(t_0), \quad \Psi^{(n+1)}(t_0)$$

is different from zero. Therefore there are exactly $2n + 2$ conditions to be satisfied in order that two twisted curves may have contact of order n.

By considering two curves whose equations are given in the form

$$y = f_1(x), \quad z = f_2(x),$$

$$y = g_1(x), \quad z = g_2(x),$$

we can show that the conditions of contact are symmetrical with respect to the two curves; for the conditions for contact at least of order n at x_0 are

$$(4·712) \quad \begin{cases} f_1(x_0) = g_1(x_0), f_1'(x_0) = g_1'(x_0), \ldots, f_1^{(n)}(x_0) = g_1^{(n)}(x_0), \\ f_2(x_0) = g_2(x_0), f_2'(x_0) = g_2'(x_0), \ldots, f_2^{(n)}(x_0) = g_2^{(n)}(x_0). \end{cases}$$

These conditions may be stated thus+: The conditions for contact at least of order n at x_0 are that y, z, and their first n differential

---

* P must of course be an ordinary point.

† There is an obvious case of exception. We must naturally suppose that the tangent to the curves is not perpendicular to the x-axis.

coefficients with respect to x should be the same at x_0 for both curves*.

Given a twisted curve depending on $2n+2$ parameters, we can define the osculating curve to a given curve at a given point, and prove that it is *in general the limit of curves of its type which intersect the given curve at P and n neighbouring points.*

Given a twisted curve depending on $2n+1$ parameters, we can in general only satisfy the conditions for contact of order $n-1$. We then have a family of curves with contact of order $n-1$, but no curve with contact of order n.

EXAMPLES III

(1) The locus of the foci of parabolas which have second order contact at a given point of a given curve is a circle.

[From Ex. II 6 deduce that, if ρ is the radius of curvature at any point of a parabola, a circle of radius $\frac{1}{4}\rho$ touching the parabola at that point passes through the focus. All the parabolas and the given curve have the same circle of curvature. Hence deduce the theorem.]

(2) Find the locus of the centres of spheres having second order contact at a given point of a given curve.

[The locus is a line parallel to the binormal through the centre of curvature. Use the results of § 3·620 to identify this locus, choosing the fundamental triad as axes of reference.]

(3) If a surface S touches a plane P along a curve C, the tangent to C at any point has third order contact with S.

[If C is $y=f(x)$, $z=0$, the surface S in the neighbourhood of $z=0$ can be put in the form $z=F\{y-f(x)\}$, $F(0)=F'(0)=0$, and we may suppose that $f(0)=f'(0)=0$. The tangent to C at the origin is $x=t$, $y=0$, $z=0$, and
$$\Phi(t)=0-F\{0-f(t)\}=-\tfrac{1}{2}F''(0)\{f(t)\}^2+O(t^6).]$$

(4) At each point M of a surface S, and through each tangent line to the surface at M, passes one (and only one) circle which has third order contact with S. Show that if M is not an umbilic there are in general ten circles which have fourth order contact with S at M.

[Darboux, *Bulletin des Sciences Math.*, ser. 2, tome IV, p. 348.]

(5) Define the osculating plane (after chapter IV) to a given curve at a given point. Show that it is unique, has in general second order contact with the curve, and is crossed by the curve. Show that this definition is consistent with previous definitions, *e.g.* that of Ex. I 9.

* For the case of two curves both given in parametric form, see Picard, Vol. I, p. 359.

An osculating plane having contact of a higher order is called *stationary*; find the condition for a stationary osculating plane, and show that if all the osculating planes are stationary the curve is a plane curve.

[d.l.V. P., vol. I, p. 335, vol. II, pp. 405 and 221.]

(6) Define the osculating line and the osculating circle to a twisted curve and prove that they are the tangent and circle of curvature respectively, having in general contact of the first and second orders with the given curve.

(7) If two curves have contact of order n at P, and Q, Q' are two points, one on each curve, such that $QQ' \gtrless (PQ)^{n+1}$, show that $PQ/PQ' \to 1$ as $Q \to P$.

[Picard, vol. I, pp. 342, 359.]

(8) Find the conditions for contact of order n for the curves

$$r = f(\theta), \quad r = g(\theta),$$

and use these conditions to obtain the tangent and the circle of curvature at any point of the curve.

CHAPTER V

THE THEORY OF ENVELOPES*

§ 5·10. **The definition of the envelope of a family of plane curves.** Consider a family of curves depending on one parameter, *viz.*

$$(5·101) \qquad\qquad f(x, y, a) = 0,$$

where $f(x, y, a)$ has as many continuous differential coefficients with respect to x, y, and a as may be mentioned : usually the first two orders will be sufficient. We suppose further that any singular points on any curve (a)† that may exist, *i.e.* points (x, y) satisfying $f = f_x' = f_y' = 0$, are isolated points. We proceed to investigate the way in which the curve (a) is placed with respect to a "neighbouring" curve of the system, *i.e.* the curve $(a + \delta a)$, where δa is sufficiently small.

Let M be a point (x, y) on the curve (a) which is an *ordinary point* of the curve, so that, for these values of x, y, and a, at least one of f_x', f_y' is not zero. It follows that a region of values of x, y, and a can be determined including M, and containing no singular point of any

---

* d.l.V.P., Vol. II, p. 408.

† *I.e.* the curve $f(x, y, a) = 0$, for which the parameter has a particular value a.

admissible member of the family*. The equation of the neighbouring curve is $f(x, y, a + \delta a) = 0$, which can be written in the form

$$(5 \cdot 102) \qquad f(x, y, a) + \delta a f_a' + O(\delta a)^2 = 0.$$

The shortest distance σ of M from this curve satisfies the relation†

$$\sigma = A f(x, y, a + \delta a)(1 + \epsilon),$$

where $A \neq 0$, (x, y) are the coordinates of M, and $\epsilon \to 0$ when $\sigma \to 0$. Since M is an ordinary point, it is easy to see that $\sigma \to 0$ when $\delta a \to 0$, so that $\epsilon \to 0$ when $\delta a \to 0$. Therefore, as $\delta a \to 0$,

$$\sigma = \{A f_a' \delta a + O(\delta a)^2\}\{1 + o(1)\}.$$

This distance will be of the second or higher order of smallness if and only if
$$f_a' = 0.$$

An ordinary point on the curve (a) whose distance from the curve $(a + \delta a)$ is of the second order of smallness at least is called **a characteristic point of the curve** (a). These points are *ordinary* points at which $f_a' = 0$. It may happen‡ that a curve (a) is entirely composed of characteristic points, but in general they will be isolated.

DEFINITION. **The envelope** *of the family*

$$f(x, y, a) = 0$$

is the locus of its isolated characteristic points.

If there is an envelope, its points satisfy the equations

$$(5 \cdot 11) \qquad f = f_a' = 0,$$

and the equation of the envelope must therefore be sought for by eliminating a between these two equations, or by solving them for x and y in terms of a and so obtaining the parametric representation

$$x = x(a), \quad y = y(a).$$

The complete result of such an elimination or solution is called **the a-discriminant of** f. We cannot, however, be certain that any curve contained in the a-discriminant forms part of the envelope. For instance, it may be possible to satisfy both equations $5 \cdot 11$ by a value of a independent of x and y, and then the corresponding curve (a) will be

---

* *I.e.* there exists a δ such that if

$$|\xi - x| \leqslant \delta, \quad |\eta - y| \leqslant \delta, \quad |A - a| \leqslant \delta,$$

it is never true that $f_\xi'(\xi, \eta, A) = f_\eta'(\xi, \eta, A) = 0$. In this region of values of (x, y), no curve of the family whose parameter satisfies $|A - a| \leqslant \delta$ has a singular point.

† See Theorem $4 \cdot 12$.

‡ This and similar statements are illustrated by examples in § $5 \cdot 20$.

composed entirely of characteristic points. Such a curve will be included in the eliminant of 5·11, and so is part of the a-discriminant, but not part of the envelope. Again, suppose that the family contains a locus of singular points given by $x = x\,(a)$, $y = y\,(a)$, the point on this locus corresponding to a being a singular point on the curve (a). The coordinates (x, y) of these singular points satisfy the equations

$$f = 0, \ f_x' = 0, \ f_y' = 0.$$

They also satisfy the equation obtained by differentiating the first equation with respect to a, *viz.*

$$f_x' \frac{dx}{da} + f_y' \frac{dy}{da} + f_a' = 0,$$

where dx/da and dy/da are determined from the equations of the singular point locus. Therefore at all points of this locus,

$$f_a' = 0.$$

Hence any locus* of singular points of curves of the family will be part of the a-discriminant, and must be distinguished from the envelope. When, however, these two classes of curves have been identified and discarded, any remaining curve or curves obtained in the foregoing manner constitute the envelope of the system.

§ 5·20. **Examples.** (i) Consider the family

$$a^2 f + (2a + 1)\, h = 0,$$

where $f = 0$, $h = 0$ are the equations of any distinct regular curves. The equations to be satisfied by a characteristic point are

$$a^2 f + (2a + 1)\, h = 0,$$
$$af + h = 0,$$

which are satisfied by every point on the curves $a = 0$, which curve is $h = 0$, and $a = -1$, which curve is $h = f$. The eliminant is

$$h\,(h - f) = 0,$$

so that in this case there is no envelope.

---

* In general no such locus exists. Singular points of members of the family are all the points determined by the equations

$$f = f_x' = f_y' = 0.$$

We have assumed that such points occur for isolated values of x and y for any given value of a. In general they will also occur only for isolated values of a, *i.e.* for isolated members of the family. It is easy to see by constructing examples that such isolated singular points *may or may not* lie on the a-discriminant.

(ii) Consider the family

$$f \equiv (y-a)^2 - x^3 = 0.$$

The curve (a) has a cusp at $(0, a)$ and therefore the $x = 0$ is a locus of singular points. But

$$f_a' \equiv -2(y-a) = 0,$$

so that the eliminant is

$$x^3 = 0,$$

which is the cusp locus, and therefore not an envelope. There is in fact no envelope.

§ 5·30. **Properties of the envelope.** Suppose that the curve (a) is not entirely composed of characteristic points. In this (the general) case, the equations $f = f_a' = 0$ determine a number of isolated points (ξ, η) on the curve (a) which are in general characteristic, but may be singular points. Let (x, y) be any point of intersection of the curves (a) and $(a + \delta a)$; then x and y are determined by the equations

(5·301) $$f(x, y, a) = 0, \quad f(x, y, a + \delta a) = 0.$$

By an application of Rolle's theorem, these may be reduced to the equations

(5·302) $$f(x, y, a) = 0, \quad f_a'(x, y, a + \mu) = 0,$$

where

$$0 < |\mu| < |\delta a|.$$

When $\mu = 0$ the solutions of equations 5·302 are the points (ξ, η) specified above. In order, therefore, to prove that the limit, as $\delta a \to 0$, of any point of intersection (x, y) is a characteristic or singular point of $f = 0$, it is only necessary to show that any solution of 5·302 determining x and y as functions of μ, is such that $x \to \xi$ and $y \to \eta$ as $\mu \to 0$. This necessary fact follows at once from implicit function theorems* if the Jacobian J of the system does not vanish, *i.e.* if

$$J = \begin{vmatrix} f_x' & f_{ax}'' \\ f_y' & f_{ay}'' \end{vmatrix} \neq 0,$$

when the variables (x, y, μ) take the system of values $(\xi, \eta, 0)$†. Even if, however, $J = 0$, it still remains true that $x \to \xi$ and $y \to \eta$ as $\mu \to 0$. Consider for example a characteristic point (ξ, η) at which $J = 0$. We may suppose without loss of generality that an arc of the curve $f = 0$ including this point can be put in the form $y = g(x)$. Substituting this value of y in f_a' we see that possible values of x must satisfy the equation

(5·303) $$h(x, \mu) \equiv f_a'\{x, g(x), a + \mu\} = 0,$$

---

* Goursat, Vol. i, Chap. iii, pp. 96, 97.

† The geometrical meaning of $J \neq 0$ is that (ξ, η) is not a singular point of $f = 0$, or of $f_a' = 0$, and that the curves $f = 0$ and $f_a' = 0$ do not touch at (ξ, η).

where $h(x, 0)$ vanishes for $x = \xi$. It may be verified that the condition $J = 0$ is equivalent to $h_x' = 0$. The point $(\xi, 0)$ may perhaps be a singular point on the (x, μ) curve*. But whether it is or not, $h(x, \mu)$ vanishes by hypothesis for values of μ in the neighbourhood of $\mu = 0$. We can, therefore, by the arguments of chapter VI, prove that there are one or more real branches of the (x, μ) curve in the neighbourhood of $(\xi, 0)$ such that

$$x = \xi + \theta_1(\mu),$$

where $\theta_1(\mu) \to 0$ as $\mu \to 0$. The corresponding value of y takes the form

$$y = \eta + \theta_2(\mu),$$

where $\theta_2(\mu) \to 0$ as $\mu \to 0$, and $\theta_2(\mu)$ is uniquely determinate when $\theta_1(\mu)$ is known. The proposition follows as before.

We have so far excluded singular points of $f = 0$. An examination of Example (ii) above shows that the limit of a point of intersection may in fact be a singular point of $f = 0$. Since, however, singular point loci are excluded from the envelope, we need not discuss this case further, admitting that it actually occurs. We have proved therefore that *the limit of any point of intersection of neighbouring curves is an isolated characteristic (or perhaps singular) point.*

The converse of this proposition is not always true. We shall content ourselves with proving it with the help of the explicit assumptions that $J \neq 0$, and $f_{aa}'' \neq 0$†. We wish to prove that if (ξ, η) is such an isolated characteristic point on the curve $f = 0$, and (x, y) a neighbouring point on the curve, then a value of δa can be found such that

$$f(x, y, a + \delta a) = 0,$$

and $\delta a \to 0$ as $x \to \xi$ and $y \to \eta$ along $f(x, y, a) = 0$. Since $J \neq 0$, there exists a unique solution of the equations

$$f(x, y, a) = 0, \quad f_a'(x, y, a + \mu) = 0,$$

near (ξ, η), such that

$$x = \xi + \theta_1(\mu), \quad y = \eta + \theta_2(\mu),$$

where $\theta_1(\mu) \to 0$ and $\theta_2(\mu) \to 0$ as $\mu \to 0$. Since $f_{aa}'' \neq 0$, $\theta_1(\mu)$ and $\theta_2(\mu)$ are not identically zero. Given a point (x, y) near (ξ, η) on $f = 0$ we can therefore find a unique number μ_1, near 0, such that

$$f_a'(x, y, a + \mu_1) = 0.$$

The point will be a singular point when in addition
$$h_\mu' \equiv f_{aa}'' = 0.$$

† When $J = 0$ for all values of a, there may be an envelope in spite of the fact that neighbouring curves do not intersect. See § 5·50. We call points at which $J \neq 0$ and $f_{aa}'' \neq 0$ *completely ordinary points*. See § 5·310.

By an application of Taylor's theorem $f_a'(x, y, a + \mu)$ can be cast into the form

$$(5\cdot304) \qquad a(x - \xi)(1 + \epsilon_1) - b\mu(1 + \epsilon_2).$$

In $5\cdot304$, y has been eliminated by using $f = 0$; a can only vanish with J and therefore $a \neq 0$; b can only vanish with f_{aa}'' and therefore $b \neq 0$; and ϵ_1, $\epsilon_2 \to 0$ as $x \to \xi$ and $\mu \to 0$.

Now since $\qquad f(x, y, a) = 0,$

$$f(x, y, a + \mu_2) = \int_0^{\mu_2} f_a'(x, y, a + \mu)\, d\mu$$

$$= a\mu_2(x - \xi)(1 + \epsilon_1) - \tfrac{1}{2} b\mu_2^2(1 + \epsilon_2).$$

Since $5\cdot304$ must vanish for $\mu = \mu_1$, $a(x - \xi)$ and $b\mu$ will have the same sign, which may be supposed positive. It follows that, as μ_1 is unique, $f_a'(x, y, a + \mu)$ is positive when $0 \leqslant \mu < \mu_1$ and vanishes *and changes sign* at μ_1. Therefore $f(x, y, a + \mu_1) > 0$, while by taking μ_2 sufficiently large we can ensure that $f(x, y, a + \mu_2) < 0$. At the same time the requisite value of μ_2 can be made as small as we please by suitably diminishing $|x - \xi|$. There exists therefore a value of δa, such that

$$f(x, y, a + \delta a) = 0, \quad (|\mu_1| < |\delta a| < |\mu_2|);$$

moreover $\delta a \to 0$ as $x \to \xi$, $y \to \eta$, and therefore the proposition is proved. We summarize this discussion in the following theorem.

THEOREM $5\cdot31$. *The locus of the limits of the intersections of neighbouring curves is in general the envelope, but may be a locus of singular points. Conversely, the envelope is in general $(J \neq 0,\ f_{aa}'' \neq 0)$ the locus of the limits of these intersections, but there may be an envelope when neighbouring curves do not intersect.*

It should be observed that "the locus of the limits of the intersections of neighbouring curves which are not singular points" is not suitable for a definition of the envelope of the system, for, as is shown in § $5\cdot50$, on that definition a curve is not the envelope of its circles of curvature.

§ $5\cdot310$. **Properties of the envelope continued. Contact with members of the family.** We now proceed to consider the behaviour of the envelope in relation to members of the family at ordinary points of the envelope. It is necessary to start with a warning. It is natural for such arguments to take any arc of the envelope as expressed in the form

$$x = x(a), \quad y = y(a),$$

where a is the parameter of the point on the envelope corresponding to the curve (a). In order that we may do so without detailed investigations, it is necessary to apply the Implicit Function Theorem to the equations

$$f(x, y, a) = 0, \quad f_a'(x, y, a) = 0,$$

so as to express x and y as functions of the parameter a.

This is possible as above* if

$$(5\cdot311) \qquad J = \begin{vmatrix} f_x' & f_{ax}'' \\ f_y' & f_{ay}'' \end{vmatrix} \neq 0$$

for the system of values, (x_0, y_0, a_0) say, in the neighbourhood of which we wish to discuss the behaviour of the envelope. The behaviour of the system in the neighbourhood of an isolated point at which $J = 0$ is very interesting geometrically, and we shall return to this case later. When $J = 0$, we cannot assert without further investigation that the envelope can be represented by $x = x(a)$, $y = y(a)$, where $x(a)$ and $y(a)$ have differential coefficients at $a = a_0$. In general we shall find that in such a case $x'(a)$ and $y'(a)$ tend to infinity as $a \to a_0$. We therefore assume for the present $J \neq 0$.

Let M be a point on the envelope, of coordinates (x, y) and parameter a, at which $J \neq 0$. It is therefore an ordinary point on the curve (a). Then in the neighbourhood of this point the envelope can be put in the form

$$x = x(a), \quad y = y(a),$$

where $x'(a)$, $y'(a)$ exist and are continuous.

The coordinates of M are functions of a which satisfy identically

$$f(x, y, a) = 0, \quad f_a'(x, y, a) = 0.$$

On differentiating with respect to a, we find that they also satisfy

$$x'(a)f_x' + y'(a)f_y' + f_a' = 0,$$
$$x'(a)f_{ax}'' + y'(a)f_{ay}'' + f_{aa}'' = 0.$$

Now $f_a' = 0$, and therefore, since $J \neq 0$, the necessary and sufficient condition that $x'(a)$ and $y'(a)$ should not be simultaneously zero is $f_{aa}'' \neq 0$.

Let us therefore suppose that M is a point on the envelope at which $J \neq 0$ and $f_{aa}'' \neq 0$. Such a point may be called a *completely ordinary point*; and in general all points will be such with the exception of isolated points. At M

$$x'(a)f_x' + y'(a)f_y' = 0,$$

* Goursat, *loc. cit.*

but as neither both f_x' and f_y' are zero, nor both $x'(a)$ and $y'(a)$, it follows that the tangents to the curve and the envelope are parallel, *i.e.* identical. Therefore the envelope touches the curve (a) at M.

Conversely, let E be a curve which at every point* touches a member of the family. Then E is a locus of points of contact with members of the family, and therefore, assuming that in general only one member of the family touches E at a given point, the coordinates (x, y) of a point of E may be expressed as functions of a, $x(a)$, $y(a)$ say, which satisfy identically

$$f(x, y, a) = 0.$$

By definition, the curve E has everywhere a tangent, but we cannot therefore assert that $x'(a)$ and $y'(a)$ must in general exist and be not both zero; and in fact no obvious method of proof on these lines presents itself. We therefore abandon the symmetry of the parametric representation and proceed as follows.

Let (x_0, y_0) be any point on E at which E is touched by the unique member of the family (a_0). Let (x, y) be a neighbouring point on E at which E is touched by (a). We have therefore

$$f(x_0, y_0, a_0) = 0, \quad f(x, y, a) = 0.$$

Now let $(x, y) \to (x_0, y_0)$. Since $f(x, y, a)$ is a continuous function of the variables,

$$f(x, y, a) = f(x_0, y_0, a) + o(1).$$

Therefore, as $(x, y) \to (x_0, y_0)$,

$$f(x_0, y_0, a) \to 0.$$

If a_0 is the only root of the equation $f(x_0, y_0, a) = 0$, then, since $f(x_0, y_0, a)$ is a continuous function of a, it follows that $a \to a_0$ as $(x, y) \to (x_0, y_0)$. In other words, neighbouring points on the curve E correspond to neighbouring values of a. This result however still remains true when the root a_0 is not unique, provided that, as we have assumed, the curve (a_0) is the only member of the family which touches E at (x_0, y_0).

Suppose for simplicity that there is only one other root a_1, so that $f(x_0, y_0, a_1) = 0$, but the curve (a_1) does not touch E at (x_0, y_0). The argument can be extended at once to the case of any finite number of other roots, so that there is no loss of generality. In this case we can establish as before that, as $(x, y) \to (x_0, y_0)$, *either* (1) $a \to a_0$, *or* (2) $a \to a_1$, *or* (3) *there exists an infinite sequence of values of a for*

---

* With the possible exception of isolated points.

which $a \to a_1$, *while the remaining values are such that* $a \to a_0$. Cases (2) and (3) may be ruled out by the following arguments. In the neighbourhood of (x_0, y_0), E is a curve having everywhere a tangent. Since this tangent is always also the tangent at (x, y) to the curve $f(x, y, a)$, for values of a in the neighbourhood of a_0 and a_1, we may suppose without loss of generality that it is never parallel to the axis of y. Hence we may suppose that E is represented by an equation of the form $y = g(x)$, where $g'(x)$ exists at all points of the interval. Let m_0 and m_1 be the slopes of $f(x, y, a_0) = 0$ and $f(x, y, a_1) = 0$ at (x_0, y_0). Then $m_0 \neq m_1$. Moreover $g'(x_0) = m_0$ and, in cases (2) and (3), $g'(x)$ assumes a series of values tending to m_1 for values of x in the neighbourhood of x_0. This however is impossible by a theorem due to Darboux[*], which states that, *if* $f(x)$ *has a differential coefficient at all points of an interval* (a, b), $f'(x)$ *cannot pass from one value to another in this interval without assuming every intermediate value.* It follows that $a \to a_0$ in all cases, which is what we required to prove.

Now consider the points (x, y, a) and $(x + \delta x, y + \delta y, a + \delta a)$ on E. Then E has a tangent at (x, y) which may without loss of generality be supposed not parallel to the axis of y.

Since $f(x, y, a) = 0$, $f(x + \delta x, y + \delta y, a + \delta a) = 0$, we have

$$\delta x f_x'(\theta) + \delta y f_y'(\theta) + \delta a f_a'(\theta) = 0,$$

where $f_x'(\theta), \ldots$, denote $f_x'(x + \theta \delta x, y + \theta \delta y, a + \theta \delta a), \ldots$, and $0 < \theta < 1$. Now let $\delta x, \delta y \to 0$ along E. We have

$$f_x'(\theta) + \frac{\delta y}{\delta x} f_y'(\theta) = - \frac{\delta a}{\delta x} f_a'(\theta).$$

But
$$f_x'(\theta) + \frac{\delta y}{\delta x} f_y'(\theta) \to f_x' + y' f_y',$$

where y' refers to the curve E. The limit y' exists because E has a tangent at (x, y). Since E touches $f(x, y, a) = 0$ at (x, y), $f_x' + y' f_y' = 0$. Also $f_a'(\theta) \to f_a'$. Therefore either $f_a' = 0$, or $\delta a / \delta x \to da/dx = 0$. The latter alternative may happen at isolated points, but cannot happen everywhere in an interval unless the curve E is identical with a member of the family. Hence in general $f_a' = 0$.

If follows that E is a locus of isolated characteristic points or of singular points, and this latter case may actually occur[†]. We have therefore proved the following theorem.

---

[*] d.l.V.P., Vol. I, p. 97.

[†] The family of curves $y^2 = (x - a)^3$ have cusps at $(a, 0)$. The line $y = 0$ touches all members of the family and touches one and only one at every point, and is also the cusp locus.

THEOREM 5·312. *The envelope touches the curve (a) at all its isolated characteristic points that are completely ordinary points of the envelope.*

Conversely, a curve which touches just one member of the family at every point is in general the envelope (or a part of it), but may be a locus of singular points.

This theorem supplies a possible alternative definition of the envelope, as "the most complete curve that touches a member of the family at every point, and which does not contain a locus of singular points". The theory is developed from this point of view by Goursat (Vol. I, p. 511); owing however to the difficulty of avoiding *a priori* assumptions as to the nature of $x'(a)$ and $y'(a)$, in the proof that on the envelope $f_a'=0$, this definition does not appear to be so suitable as the one chosen here.

A corollary of this last theorem is that any curve is the envelope of its tangents. A direct proof is not without interest. Let $y=f(x)$ be the given curve, so that the tangent at the point a is

$$y - f(a) - (x-a)f'(a) = 0,$$

and the coordinates of the characteristic points satisfy

$$(x-a)f''(a) = 0.$$

If $f''(a)=0$, every point on the tangent, which is then inflexional, is a characteristic point: discarding this case, the isolated characteristic point of the tangent at a is the point $x=a$, $y=f(a)$, and the locus of these points is the given curve.

§ 5·40. Order of contact of the envelope and the curves. Meaning of $J=0$.

We have already proved that, at any completely ordinary point a of the envelope, the envelope touches the curve (a). It is easily seen that the contact is necessarily first order contact at such a point, for $J=0$ is a necessary condition for contact of higher order than the first*.

We now consider a point a on the envelope which is an ordinary point of the curve (a), but at which $J=0$; in general $f_{aa}''\neq0$ at such a point; we shall suppose that this is the case. Near such a point we cannot assume at once that the parametric representation of the envelope $x=x(a)$, $y=y(a)$ is possible, where $x(a)$ and $y(a)$ possess differential coefficients. We can prove however by differentiating along the envelope, exactly as in the latter part of Theorem 5·312, *that $a'=da/dx$ exists at this point and, because $J=0$, is in fact zero, and that the envelope touches the curve (a).*

One aspect of the geometrical meaning of the conditions $J=0$, $f_{aa}''\neq0$ at an isolated point of the envelope is therefore that *the distribution of points of contact of members of the family along the envelope is exceptionally sparse near such a point.*

We can however go further than this. We can prove step by step, with suitable assumptions as to the nature of $f(x, y, a)$, that a'', a''', ... exist when

---

* A proof is sketched in Ex. IV, 3.

we differentiate along the envelope. Now the envelope satisfies identically $f(x, y, a)=0$ and $f_a'(x, y, a)=0$, and therefore satisfies identically

$$(5\text{·}401) \qquad\qquad f_x' + y' f_y' + a' f_a' = 0,$$

$$(5\text{·}402) \qquad\qquad f_{ax}'' + y' f_{ay}'' + a' f_{aa}'' = 0.$$

Differentiating 5·401 again along the envelope, and using 5·402 to simplify the result, we obtain

$$(5\text{·}403) \qquad f_{xx}'' + 2y' f_{xy}'' + y'^2 f_{yy}'' + y'' f_y' - a'^2 f_{aa}'' + a'' f_a' = 0.$$

Differentiating 5·403, and using the facts that $a'=0$ at the point a, and that, as always, $f_a'=0$, we obtain

$$(5\text{·}404) \quad f_{xxx}''' + 3y' f_{xxy}''' + 3y'^2 f_{xyy}''' + y'^3 f_{yyy}'''$$
$$+ 3y'' f_{xy}'' + 3y' y'' f_{yy}'' + y''' f_y' = 0.$$

Equation 5·403 itself reduces to

$$(5\text{·}405) \qquad\qquad f_{xx}'' + 2y' f_{xy}'' + y'^2 f_{yy}'' + y'' f_y' = 0.$$

But 5·404 and 5·405 are precisely the expressions we obtain when we determine y'' and y''' for the curve (a) by differentiating $f(x, y, a)=0$ with a constant. It is easily verified that in general y^{iv} is different for the curve and the envelope. At this point therefore the curve (a) and the envelope have in general contact of the third order, i.e. *two orders higher than normal*. We can collect these results into the following theorem.

THEOREM 5·41. *At isolated points a on the envelope at which $J=0$, $f_{aa}'' \neq 0$, and the curve (a) has an ordinary point,*

   (i) *the envelope has in general third order contact with the curve (a),*

   (ii) *$da/dx = da/dy = 0$*, so that the distribution of points of contact of members of the family along the envelope is exceptionally sparse†.*

The condition $f_{aa}''=0$, $J \neq 0$, is in general satisfied at isolated points of the envelope. On referring to § 5·310, we see that this condition implies that $x'(a)=y'(a)=0$, so that the point is a singular point on the envelope. It is in fact in general a cusp of the first species (see Chapter VI).

If both the conditions $J=0$ and $f_{aa}''=0$ are satisfied at an isolated point of the envelope, the state of affairs is more complicated. In general the envelope has two branches through the point, both of which have second order contact with the curve (a). In certain cases one of these branches may coincide with the curve (a)‡. It is then no longer part of the envelope, but still remains part of the a-discriminant. For the further study of these and other singularities of the envelope, or more generally of the a-discriminant, the reader should refer to Bromwich and Hudson, *loc. cit.*

---

   * Assuming that the tangent to the curve or envelope is parallel to neither axis of coordinates.

   † In this connection the reader should refer to a paper by Bromwich and Hudson (*Quarterly Journal*, Vol. XXXIII, p. 98) called "The discriminant of a family of curves or surfaces."

   ‡ An inflexional tangent is the simplest example. The reader should verify that $J=0$ and $f_{aa}''=0$ in this case. See p. 67.

§ 5·420. **Envelopes with contact everywhere of high order.** We have so far considered the behaviour of the envelope at isolated exceptional points. It may happen however that every point is exceptional, so that $J = 0$ for all values of a. By definition, no arc of the envelope can coincide with any member of the family, and therefore a' can only vanish at isolated points. We have as usual, differentiating along the envelope,

$$f_x' + y' f_y' = 0,$$

$$f_{ax}'' + y' f_{ay}'' + a' f_{aa}'' = 0,$$

so that $J = 0$ *for all values of a if and only if $f_{aa}'' = 0$ for all values of a.*

In general $f = 0$ and $f_{aa}'' = 0$ will determine (by the implicit function theorem) a parametric representation of the envelope, $x = x(a)$, $y = y(a)$, possessing differential coefficients near any value of a except those isolated values at which

$$(5·421) \qquad J_1 = \begin{vmatrix} f_x' & f_{aax}''' \\ f_y' & f_{aay}''' \end{vmatrix} = 0.$$

When $J_1 \neq 0$, the point is an ordinary point on the envelope unless $f_{aaa}''' = 0$.

The conditions for contact of order n between the envelope and the curve (a) are, by 4·212,

$$(5·422) \qquad \left[\frac{\partial^r \Phi(t, a)}{\partial t^r}\right]_{t=a} = 0, \quad (r = 1, 2, \dots n),$$

$$(5·423) \qquad \left[\frac{\partial^{n+1} \Phi(t, a)}{\partial t^{n+1}}\right]_{t=a} \neq 0,$$

where

$$\Phi(t, a) = f(x(t), y(t), a).$$

We observe that the equations

$$(5·424) \qquad \Phi(t, t) = 0, \quad \Phi_a'(t, t) = 0, \quad \Phi_{aa}''(t, t) = 0$$

are satisfied for all values of t, where $\Phi_a'(t, t)$ denotes

$$[f_a'(x(t), y(t), a)]_{a=t}.$$

Making use of the identities 5·424 and the similar identities obtained by differentiation, we verify that 5·422 and 5·423 are satisfied for $n = 2$. The envelope therefore has in general second order contact with all members of the family. At isolated points at which $J_1 = 0$, $f_{aaa}''' \neq 0$, we can show as in the last section that $a' = 0$ and the envelope has in general third order contact with the curve (a). In this case the contact is only one order higher than usual.

Conversely, if the contact is in general second order with all members of the family, then $J = 0$ and $f_{aa}'' = 0$ for all values of a, for otherwise contact is in general of the first order only. It therefore follows that the necessary and sufficient conditions that the envelope should in general have contact of the second order with all members of the family are that

$$f_{aa}'' = 0$$

for all values of a, and

$$f_{aaa}''' \neq 0$$

except for isolated values of a.

The same arguments can be extended step by step to prove that the necessary and sufficient conditions that the envelope should in general have contact of order n with all members of the family are that

$$(5\text{·}425) \qquad \frac{\partial^2 f}{\partial a^2} = \frac{\partial^3 f}{\partial a^3} = \ldots = \frac{\partial^n f}{\partial a^n} = 0$$

for all values of a, and

$$(5\text{·}426) \qquad \frac{\partial^{n+1} f}{\partial a^{n+1}} \neq 0$$

except for isolated values of a.

It should be observed that these conditions 5·425 and 5·426 bear no obvious relation whatever to the necessary and sufficient conditions for contact of order n at an isolated point of an envelope for which contact is in general of order $n - r$, where $r \geqslant 1$*.

In accordance with Theorem 4·22, the members of the family in general cross the envelope at their points of contact when n, the order of the contact, is even, and do not cross the envelope when n is odd. It can also be shown that in a family for which n is in general odd, neighbouring curves must intersect, while in a family for which n is in general even neighbouring curves do not intersect†. To prove these assertions an extension of the analysis of § 5·30 is required to the more complicated cases in which $J = 0$. We shall content ourselves here with proving that, when $n = 2$, neighbouring curves do not intersect. In this case we have $J = 0$, $f_{aa}'' = 0$ everywhere and in general $f_{aaa}''' \neq 0$.

The proof of § 5·30 that, when $n = 1$, neighbouring curves must intersect depends essentially on the fact that, if (x, y) is any given point on

---

* It is stated by Goursat (Vol. i, p. 549) in an example, that if 5·425 and 5·426 are satisfied for an *isolated* value a, they are the conditions for a contact of order n between the envelope and the curve (a). This is incorrect.

† This is geometrically obvious, or almost so, as can be seen by drawing a figure.

the curve (a) near a characteristic point, a value μ_1 of μ can be found, near 0, such that

$$f_a'\,(x,\,y,\,a+\mu_1)$$

vanishes and *changes sign* when μ passes through the value μ_1. The failure of the argument for $n=2$ (and in general when n is even) is due to the fact that the lowest order terms in $f_a'\,(x,\,y,\,a+\mu)$ (except for a constant factor) form a perfect square, and no value of μ exists for which f_a' changes sign near $\mu=0$. Taking for simplicity the point $(0,\,0)$ as the characteristic point under discussion on the curve (a), it is easily verified that the lowest order terms in the expansion of $f_a'\,(x,\,y,\,a+\mu)$, near $(0,\,0)$ and near $\mu=0$, are second order terms in x and μ, which can be put in the form

$$\frac{1}{2\,!}\{x^2\,(f_{axx}{}''' + 2y'f_{axy}{}''' + y'^2 f_{ayy}{}''' + y''f_{ay}{}'')_0$$
$$+\,2x\mu\,(f_{aax}{}''' + y'f_{aay}{}''')_0 + \mu^2\,(f_{aaa}{}''')_0\},$$

where y' and y'' refer to the curve (a). But both y' and y'' in this case have the same values for the envelope at $(0,\,0)$; also at all points of the envelope $f_a' = f_{aa}'' = 0$. By differentiating these equations along the envelope we obtain finally

$$(f_{axx}{}''' + 2y'f_{axy}{}''' + y'^2 f_{ayy}{}''' + y''f_{ay}{}'')_0 = a_0'^2\,(f_{aaa}{}''')_0,$$
$$(f'''_{aax} + y'f'''_{aay})_0 = -\,a_0'\,(f'''_{aaa})_0,$$

where $a_0' = (da/dx)_0$, taken along the envelope ; in general a_0' is not zero. The second order terms therefore reduce to

$$\frac{1}{2\,!}\,(f_{aaa}{}''')_0\,(xa_0' - \mu)^2.$$

It follows that, for any given values of x and y on the curve (a) near $(0,\,0)$, f_a' cannot change sign for any value μ such that $|\,\mu\,| < A$, where A is a constant independent of x and y. For such values of μ, $\displaystyle\int_0^\mu f_a'\,d\mu$ can never vanish. No value therefore of δa can be found for which $f\,(x,\,y,\,a+\delta a)=0$, while δa tends to zero as $(x,\,y)$ tends to a characteristic point on (a). Therefore no characteristic point is the limit of points of intersection of neighbouring curves. If such curves intersect at all they can only do so near a singular point. These results may be summarized thus :

Theorem 5·427. *Under the conditions* 5·425 *and* 5·426, *the contact between the curve and its envelope is of order* n, *except at isolated points at which the order may be higher. Neighbouring curves* (a) *in general intersect and do not cross the envelope when* n *is odd, and do not intersect but cross the envelope when* n *is even.*

72 THE THEORY OF ENVELOPES

§ 5·50. **The envelope of a system of circles.** We will now give an example of a family for which neighbouring curves do not intersect, but which still possesses an envelope. Such a family is formed by the circles of curvature of any plane curve. For since the plane curve touches each circle at an ordinary point, it is the envelope (or part of it). Moreover, on referring to § 3·70, we see that the difference between the radii of curvature at neighbouring points is equal to the arc of the evolute between the two centres of curvature, and is therefore in general greater than the distance between these centres. One circle therefore completely encloses the other, and so there are no points of intersection.

It follows from § 5·420 that contact between members of the family and the envelope must in such a case be of even order, and therefore the typical case of an envelope not generated by limits of points of intersection is that in which this contact is in general of the second order. A family of circles is the simplest possible family of the kind, for the three conditions of second order contact require just three arbitrary functions of a in the equation of the curves of the family.

It is of interest to study a family of circles directly. We shall, among others, arrive at the foregoing results, and also find that a given curve is the complete envelope of its circles of curvature, a point at present in doubt. Let the system be

$$(5·501) \qquad (x-a)^2+(y-b)^2-R^2=0$$

where a, b, R are functions of a. There are no singular points on any member. The characteristic points are the intersections of this circle with the straight line

$$(5·502) \qquad (x-a)\,a'+(y-b)\,b'+RR'=0.$$

Denoting the locus of the centres of the circles by C, we see that this line is perpendicular to the tangent to C at a, *i.e.* the point $a\,(a)$, $b\,(a)$, and distant from this point by

$$RR'/(a'^2+b'^2)^{\frac{1}{2}}\,*.$$

This is greater than, equal to, or less than R according as R'^2 is greater than, equal to, or less than $a'^2+b'^2$, and there are no, one, or two characteristic points respectively. In the first case there is no envelope for that part of the family, and, if $R'^2>a'^2+b'^2$ for all values of a, no envelope at all. Next suppose that for all values of a, $R'^2<a'^2+b'^2$. There is then an envelope, composed in general of two branches; an obvious example is provided by the case of a moving circle of constant radius $R'=0$. Finally suppose that for all values of a

$$(5·51) \qquad R'^2=a'^2+b'^2.$$

---

* Supposing a' and b' not both zero. If they both are, and $R'\neq0$, there is no characteristic point, but if also $R'=0$, this particular circle is entirely composed of characteristic points. This can only happen for isolated values of a, which may be neglected.

There is then an envelope composed of the points of contact of 5·501 and 5·502. It follows that $(x-a)\,a' + (y-b)\,b' + RR' = 0$ is the tangent to the envelope at the characteristic point of a, and therefore that normals to the envelope are tangents to C. In other words C is the evolute of the envelope, and therefore *the circles are the circles of curvature of their envelope.* It is easily seen that the condition 5·51 is also necessary for this relationship. Any plane curve is therefore the complete envelope of its circles of curvature.

§ 5·60. **Other rules for envelopes.** (1) The equation of the family may be given in some other form such as

$$(5\text{·}601) \qquad\qquad f(x, y, a, \beta) = 0$$

with the condition

$$(5\text{·}602) \qquad\qquad \phi\,(a, \beta) = 0.$$

We apply the usual rule, regarding β as a function of a, and must therefore have

$$(5\text{·}603) \qquad\qquad f_a' + \frac{d\beta}{da} f_\beta' = 0.$$

But we also have

$$(5\text{·}604) \qquad\qquad \phi_a' + \frac{d\beta}{da}\,\phi_\beta' = 0$$

and we have therefore to eliminate a, β, $\dfrac{d\beta}{da}$ between the equations (5·601—5·604), or, what is the same thing, a and β between

$$(5\text{·}61) \qquad\qquad f = 0, \quad \phi = 0, \quad \frac{\partial\,(f,\,\phi)}{\partial\,(a,\,\beta)} = 0.$$

All the former exceptional cases must be taken account of, with the addition of singularities of $\phi\,(a, \beta) = 0$, but these, being isolated points, are not of importance.

(2) The curves may be given in the parametric form

$$(5\text{·}611) \qquad\qquad x = \phi_1\,(t, a), \quad y = \phi_2\,(t, a).$$

We can apply the usual rule, regarding t as a function of y and a. We have therefore

$$\frac{\partial\phi_1}{\partial t}\frac{\partial t}{\partial a} + \frac{\partial\phi_1}{\partial a} = 0,$$

corresponding to $f_a' = 0$, and also the identical relation

$$\frac{\partial\phi_2}{\partial t}\frac{\partial t}{\partial a} + \frac{\partial\phi_2}{\partial a} = 0,$$

which lead together to

$$\frac{\partial\,(\phi_1,\,\phi_2)}{\partial\,(t,\,a)} = 0.$$

The envelope is therefore to be looked for, with the usual precautions, among the results of eliminating t and a from

$$(5\cdot62)\qquad x=\phi_1(t,a),\quad y=\phi_2(t,a),\quad \frac{\partial(\phi_1,\phi_2)}{\partial(t,a)}=0.$$

EXAMPLES IV

(1) *The envelope of the straight line*

(1) $$x\cos a+y\sin a=f(a).$$

The characteristic point lies on the straight line

(2) $$-x\sin a+y\cos a=f'(a).$$

Show that (i) the line (2) is normal to the envelope, (ii) the envelope of the line (2) is the evolute of that of the line (1), and (iii) that the curvature and arc of the envelope of the line (1) are given by

$$\rho=ds/da=\pm\{f(a)+f''(a)\}.$$

(2) *Families of circles, such that $R'^2=a'^2+b'^2$ (§ 5·50).*
The family is

$$(x-a)^2+(y-b)^2-R^2=0.$$

Calculate δR and $\{(\delta a)^2+(\delta b)^2\}^{\frac{1}{2}}$, corresponding to a positive change of δa in a, as far as terms in $(\delta a)^3$, and thus verify that neighbouring circles do not intersect.

[We find
$$\delta R=A\,\delta a+B\,(\delta a)^2+C_1\,(\delta a)^3+O\,(\delta a)^4,$$
$$\{(\delta a)^2+(\delta b)^2\}^{\frac{1}{2}}=A\,\delta a+B\,(\delta a)^2+C_2\,(\delta a)^3+O\,(\delta a)^4,$$

where
$$A=(a'^2+b'^2)^{\frac{1}{2}},\quad B=(a'a''+b'b'')/2A,$$
$$C_1=\{(\Sigma a'a'''+\Sigma a''^2)\,A^2-(\Sigma a'a'')^2\}/6A^3,$$
$$C_2=\{(\tfrac{1}{6}\Sigma a'a'''+\tfrac{1}{8}\Sigma a''^2)\,A^2-\tfrac{1}{8}(\Sigma a'a'')^2\}/A^3,$$

where Σ denotes summation over the functions a and b; so that

$$\delta R-\{(\delta a)^2+(\delta b)^2\}^{\frac{1}{2}}=\frac{\Sigma a'^2\,\Sigma a''^2-(\Sigma a'a'')^2}{24A^3}\,(\delta a)^3+O\,(\delta a)^4,$$

which is positive when δa is small.]

(3) *Curvature of the envelope.* Let y', y'' ... denote the derivatives of y along the curve (a), and y_e', y_e'' ... the derivatives along the envelope. Then

$$y'=y_e'=-f_x'/f_y',$$

and the difference of the curvatures is given by

$$\frac{1}{\rho_e}-\frac{1}{\rho}=\frac{y_e''-y''}{(1+y'^2)^{\frac{3}{2}}}.$$

Differentiating $f=0$ twice along the curve (a) we obtain

(1) $$f_{xx}''+2y'f_{xy}''+y'^2f_{yy}''+y''f_y'=0;$$

and differentiating along the envelope

(2) $$f_{xx}''+2y'f_{xy}''+y'^2f_{yy}''+y_e''f_y'+2a'f_{ax}''+2a'y'f_{ay}''+a'^2f_{aa}''+a''f_a'=0.$$

Subtracting (1) from (2) and using the facts that for the envelope

$$f_a'=0, \quad f_{ax}''+y'f_{ay}''+a'f_{aa}''=0,$$

we have

$$\frac{1}{\rho_e}-\frac{1}{\rho}=\frac{(f_x'f_{ay}''-f_y'f_{ax}'')^2}{f_{aa}''(f_x'^2+f_y'^2)^{\frac{3}{2}}}=\frac{J^2}{f_{aa}''(f_x'^2+f_y'^2)^{\frac{3}{2}}}.$$

We thus arrive as before at the fact that the curvatures cannot be equal unless $J=0$. Equality of curvature at a point of contact is equivalent to second order contact.

(4) Verify that, for the family of circles

$$(x-a)^2+(y-b)^2-R^2=0,$$

the condition $J=0$ reduces to $R'^2=a'^2+b'^2$.

(5) Show that the curvature of the curve whose tangential equation is $\phi(l, m)=0$ is

$$\frac{(l\phi_l'+m\phi_m')^2}{(l^2+m^2)^{\frac{3}{2}}(\phi_{ll}''\phi_m'^2-2\phi_{lm}''\phi_l'\phi_m'+\phi_{mm}''\phi_l'^2)}$$

at its point of contact with $lx+my=1$.

[The curve is the envelope of $lx+my=1$ under the condition $\phi(l, m)=0$. Take l as parameter and let $m_1, m_2, \ldots$ be the derivatives of m with respect to l. We have

$$m_1=-\phi_l'/\phi_m', \quad m_2=-(\phi_{ll}''\phi_m'^2-2\phi_{lm}''\phi_l'\phi_m'+\phi_{mm}''\phi_l'^2)/\phi_m'^3.$$

The coordinates of the characteristic point satisfy

$$lx+my=1, \quad x+m_1y=0.$$

Apply the general formula of Example 3, remembering that $1/\rho=0$.]

(6) If A, B, C are functions of x and y, show that in general the envelope of $Aa^2+2Ba+C=0$ is $B^2=AC$. Examine the exceptional cases. [See § 5·20.]

(7) The envelope of the family

$$\phi(x, y, a, \beta, \gamma)=0, \quad f_1(a, \beta, \gamma)=0, \quad f_2(a, \beta, \gamma)=0$$

is to be found in the result of eliminating a, β, γ from these equations and

$$\frac{\partial(\phi, f_1, f_2)}{\partial(a, \beta, \gamma)}=0.$$

(8) The first positive pedal of a plane curve is the envelope of circles described on the radii vectores as diameters. The first negative pedal is the envelope of a straight line drawn through any point of the curve and at right angles to the radius vector to the point. [Lamb, *Infinitesimal Calculus*, 2nd ed., p. 382.]

(9) *Caustics.* The caustic of a curve C with respect to a luminous point A is defined as the envelope of the rays from A after reflection by C. Prove that the caustic is the evolute of the pedal, with respect to A, of the curve C',

similar to C, which is obtained by producing each radius vector from A to the curve C a distance equal to itself. Deduce that

$$\frac{1}{l} + \frac{1}{r} = \frac{2}{R \cos i},$$

r being the length of the incident ray AP, l the length of the reflected ray from P to the caustic, R the radius of curvature of C at P, and i the angle of incidence.

In particular, if the rays of light are parallel

$$l = \tfrac{1}{2} R \cos i,$$

and the normal to the caustic passes through the middle point of the radius of curvature of C.

[The normal to the pedal of a curve passes through the middle point of the radius vector (Ex. 8). Hence the reflected ray is normal to the pedal of C' with respect to A. Use the relation between the curvatures of a curve and its pedal (d.l.V.P., Vol. I, p. 323, Ex. 8).]

§ 5·70. **Similar problems in three dimensions.** Many three-dimensional problems on envelopes are direct extensions of the two-dimensional ones already treated. We shall consider some of the simplest.

(1) *Envelope of a family of surfaces*

$$f(x, y, z, a) = 0.$$

We define a *characteristic point* exactly in § 5·10 and show that it is an ordinary point of the surface satisfying

$$(5·71) \qquad\qquad f = 0, \quad f_a' = 0.$$

For certain values of a, the whole surface may be composed of characteristic points*, but in general there will be an isolated curve of characteristic points lying on each surface $f = 0$. Such a curve is called **the characteristic of the surface** $f = 0$. We then define the envelope of the family as *the surface generated by the characteristics*. Following the lines of §§ 5·10—5·60, we then prove that the envelope is obtained by eliminating a from $f = 0$, $f_a' = 0$, taking care to discard the stationary surfaces, and surfaces composed of singular lines of the surfaces (a); that when neighbouring surfaces intersect, the limits of the curves of intersection are, in general, characteristics, but that envelopes may exist when neighbouring surfaces do not intersect; that a surface (a) touches the envelope at every point of its characteristic; and that a surface consisting of curves of contact with members of the family is in general

---

* These may be called "*stationary surfaces*," if a name is required.

the envelope (or part of it). An alternative enunciation of the last assertion is that a surface which, at every point, touches one of the surfaces of the family is, in general, the envelope (or part of it).

In particular, a surface whose tangent plane depends on one parameter, is the envelope of its tangent plane. Such a surface is, of course, a developable surface.

(2) *Envelope of a family of surfaces*

$$f(x, y, z, a, \beta) = 0.$$

We define a *characteristic point* as an ordinary point of the surface whose distance from the surface

$$f(x, y, z, a + \delta a, \beta + \delta \beta) = 0$$

is small compared to $|\delta a| + |\delta \beta|$. Such points are ordinary points satisfying

$$(5\cdot72) \qquad f = 0, \quad f_a' = 0, \quad f_\beta' = 0,$$

and are, in general, isolated, but whole surfaces or whole lines on particular surfaces may be composed of characteristic points. The surface composed of all isolated characteristic points or lines of characteristic points is called the envelope. As before, the envelope may or may not be generated by the limits of points of intersection of the surfaces

$$f(x, y, z, a, \beta) = 0,$$
$$f(x, y, z, a + \delta a, \beta) = 0,$$
$$f(x, y, z, a, \beta + \delta \beta) = 0,$$

and touches every surface of the family. Conversely, a surface touching every member of the family is, in general, the envelope (or part of it).

In particular, a surface whose tangent plane depends on two parameters (*i.e.* any surface) is the envelope of its tangent planes.

(3) *Envelope of the family of twisted curves*

$$(5\cdot73) \qquad f(x, y, z, a) = 0, \quad g(x, y, z, a) = 0.$$

We define a characteristic point of the curve (a) as an ordinary point whose distance from the curve $(a + \delta a)$ is of order higher than the first in δa. We find, by the former method, that such a point must satisfy

$$(5\cdot731) \qquad f_a' = 0, \quad g_a' = 0.$$

In general, the equations $5\cdot73$ and $5\cdot731$ are incompatible (except perhaps for a number of isolated values of a), and there are no characteristic points. But if, as may happen, the equations $5\cdot73$ and

5·731 effectively reduce to three, and so can be satisfied by the continuous functions of a

$$x = x(a), \quad y = y(a), \quad z = z(a),$$

the twisted curve, of which the above functions give the parametric representation, is called the envelope of the family. We can then prove, as before, that the envelope, if it exists, touches all the curves at their characteristic points, and, conversely, that if a curve exists which at every point touches one member of the family, then this curve is, in general, the envelope (or part of it).

There is one important particular case in which the envelope in general exists, and that is the case of the characteristics of a family of surfaces depending on one parameter. The characteristics satisfy the equations

$$f = 0, \quad f_a' = 0,$$

and therefore the four equations 5·73 and 5·731 reduce in this case to three

$$(5·74) \qquad f = 0, \quad f_a' = 0, \quad f_{aa}'' = 0,$$

which in general will just determine the isolated characteristic points whose locus is the envelope. This curve which touches all the characteristics of the family, lies on the envelope and is called its *edge of regression**.

For proofs of many of these assertions and for further developments in the region of Solid Geometry the reader is referred to de la Vallée Poussin, Picard, and the paper (already quoted) by Bromwich and Hudson.

EXAMPLES V

(1) The edge of regression has, in general, second order contact with any surface of the family on the characteristic.

[Use Theorem 4·61.]

(2) Discuss the problem of the envelope of a moving sphere, whose centre lies on a twisted curve, (i) when the radius is constant (*une surface canal*), (ii) when the radius varies.

[Extend § 5·50 from circles to spheres.]

---

* Fr. *l'arête de rebroussement*. Plane sections of the envelope have in general a cusp (*point de rebroussement*) where the plane cuts this curve. See Picard, vol. I, p. 322.

(3) *The wave surface.* The envelope of the plane

$$ax + \beta y + \gamma z = l,$$

where

$$a^2 + \beta^2 + \gamma^2 = 1$$

and

$$\frac{a^2}{l^2 - a^2} + \frac{\beta^2}{l^2 - b^2} + \frac{\gamma^2}{l^2 - c^2} = 0,$$

is the surface

$$\frac{a^2 x^2}{r^2 - a^2} + \frac{b^2 y^2}{r^2 - b^2} + \frac{c^2 z^2}{r^2 - c^2} = 0, \quad (r^2 = x^2 + y^2 + z^2).$$

(4) The direction cosines of the normal to the surface obtained by eliminating a between $f(x, y, z, a) = 0$ and $g(x, y, z, a) = 0$ are given by

$$l : m : n = \frac{\partial(f, g)}{\partial(x, a)} : \frac{\partial(f, g)}{\partial(y, a)} : \frac{\partial(f, g)}{\partial(z, a)}.$$

Obtain the corresponding result for a surface obtained by elimination of two parameters between three equations. Apply these results to the proof of the contact of a surface with an envelope.

(5) The envelope of

$$f(x, y, z, a, \beta, \gamma) = 0,$$

where

$$\phi(a, \beta, \gamma) = 0, \quad \psi(a, \beta, \gamma) = 0,$$

is given by

$$f = 0, \quad \phi = 0, \quad \psi = 0, \quad \frac{\partial(f, \phi, \psi)}{\partial(a, \beta, \gamma)} = 0.$$

(6) The envelope of

$$f(x, y, z, a, \beta, \gamma) = 0,$$

where

$$\phi(a, \beta, \gamma) = 0,$$

is given by

$$f = 0, \quad \phi = 0, \quad f_a'/\phi_a' = f_\beta'/\phi_\beta' = f_\gamma'/\phi_\gamma'.$$

(7) Given a surface S, suppose that with each point m of S as centre a sphere Σ is drawn with variable radius R. Prove that this sphere Σ in general touches its envelope in two points M and M' such that MM' is normal to the tangent plane at m to S.

CHAPTER VI

SINGULAR POINTS OF PLANE CURVES*

§ 6·10. **Exceptional cases in general. Singular points.** In the course of the preceding chapters we have encountered cases of exception in which, when some particular relation is exactly satisfied, the general treatment usually applicable breaks down. These cases of exception correspond to exceptional points on the curve, usually such that there are only a finite number in any finite region, at which the curve has some peculiar property such as a stationary tangent, a point of inflexion, exceptionally high order contact with its circle of curvature, or its envelope, etc. All such points at which the curve possesses peculiar properties may be considered to be **singular points** on the curve, but it is usual to reserve this name for a particular class of exceptional points—the most important class—which consists of the points at which

$$f_x' = f_y' = 0$$

when the curve is $f(x, y) = 0$, and

$$\phi_1'(t) = \phi_2' t_j = 0$$

when the curve is $x = \phi_1(t)$, $y = \phi_2(t)$. The former case is more general, and we shall be mainly concerned with it. For the neighbourhood of such points the general existence theorem 1·51 breaks down, and further investigation is needed.

We shall again in this chapter, as in Chapters IV and V, tacitly assume the existence and continuity, in the neighbourhood of the singular point under discussion, of all the partial differential coefficients of $f(x, y)$, or differential coefficients of $\phi_1(t)$, $\phi_2(t)$, that are mentioned.

§ 6·20. **Form of $f(x, y) = 0$ in the neighbourhood of a singular point.** Without loss of generality we may suppose that the singular point is the origin, so that

$$f(0, 0) = f_x'(0, 0) = f_y'(0, 0) = 0.$$

It follows from Taylor's Theorem that in the neighbourhood of the origin $f(x, y)$ can be put in the form

$$(6·21) \qquad f(x, y) = \phi_2 + \phi_3 + \dots + \phi_n + R_n.$$

---

* This chapter follows d.l.V.P., Vol. II, Chap. IX.

In this equation

$$\phi_n = \frac{1}{n!}\left\{x^n\left(\frac{\partial^n f}{\partial x^n}\right)_0 + {}_nC_1 x^{n-1}y\left(\frac{\partial^n f}{\partial x^{n-1}y}\right)_0 + \ldots + y^n\left(\frac{\partial^n f}{\partial y^n}\right)_0\right\}$$
$$= a_n x^n + b_n x^{n-1}y + \ldots + k_n y^n,$$

the latter form being used for the sake of shortness, and R_n is the remainder after n terms. The most useful form of R_n for our purposes is the exact form[*]

$$(6\text{·}211) \qquad R_n = \frac{x^{n+1}}{n!}\int_0^1 F^{(n+1)}(tx)(1-t)^n dt,$$

where

$$F(x) = F(x, u) = f(x, ux) \quad (u = y/x),$$

or the corresponding form

$$(6\text{·}212) \qquad R_n = \frac{y^{n+1}}{n!}\int_0^1 G^{(n+1)}(ty)(1-t)^n dt,$$

where

$$G(y) = G(y, v) = f(vy, y) \quad (v = x/y).$$

Provided that, as is generally the case, ϕ_2 does not vanish identically, the singular point is said to be *of the second order*. When ϕ_2, $\phi_3 \ldots \phi_{n-1}$ are identically zero, but ϕ_n is not, the singular point is said to be *of order n*.

The fundamental form of the curve will not be altered by any transformation of the axes of the type

$$x = ax' + by', \quad y = cx' + dy';$$

and the properties of the transformed function $f'(x', y')$ will be the same as those of $f(x, y)$ near $(0, 0)$. We may therefore at any stage make, without loss of generality, any such transformation that simplifies the discussion.

Certain properties of the transformed functions $F(x, u)$, $G(y, v)$ should be noted. The function $F(x, u)$ has as many orders of partial differential coefficients as $f(x, y)$ near $x = 0$ *and any finite value of u,* for all such points belong to the neighbourhood of $x = 0$, $y = 0$. A similar statement holds of $G(y, v)$ near $y = 0$, *and any finite value of v.* Any point near $x = 0$, $y = 0$, corresponds to a finite value of one at least of u or v.

If we suppose for example that u is finite, and apply 6·211, we see that in such a neighbourhood

$$(6\text{·}22) \qquad R_n = O\left(x^{n+1}\right).$$

---

[*] d.l.V.P., *loc. cit.*, and Vol. I, p. 432.

§ 6·30. **Nature of the curves defined by** $f(x, y) = 0$. A discussion of the following simple case will illustrate the nature of the curves defined by $f(x, y) = 0$. Let

$$f(x, y) \equiv (y - \alpha x)(y - \beta x) = 0,$$

where α and β are real constants. Then the equation $f(x, y) = 0$ defines the two straight lines

$$y = \alpha x, \quad y = \beta x.$$

These are two independent solutions of $f(x, y) = 0$, which are continuous in the neighbourhood of the singular point $(0, 0)$. These solutions, however, are not the only functions of the type $y = \phi(x)$ which satisfy $f(x, y) = 0$.

The function (or curve)

$$y = \alpha x \ (x \ rational), \quad y = \beta x \ (x \ irrational)$$

also satisfies $f(x, y) = 0$. This is a difficulty that does not occur in the non-exceptional case (Theorem 1·51). It is essentially due to the fact that in this case there are *two independent* solutions near $(0, 0)$. We see therefore that it is necessary to make the *a priori* restriction that we are concerned only with *continuous* solutions of $f(x, y) = 0$, *i.e.* such curves as can be put in one of the forms

$$y = \phi(x), \quad x = \phi(y),$$

where $\phi(x)$ or $\phi(y)$ is a continuous function *in the neighbourhood of* $x = 0$, or $y = 0$, except perhaps *at* $x = 0$, or $y = 0$ itself. There may be an infinity of other solutions of $f(x, y) = 0$, which are of no present interest.

It is not necessary to assume *a priori* that, for example, $\phi(x)$ is continuous at $x = 0$, *i.e.* $\phi(x) \to 0$ as $x \to 0$. Since we are considering solutions of $f(x, y) = 0$ in the neighbourhood of $(0, 0)$ it is only necessary *a priori* that zero should lie between the limits of indetermination of $\phi(x)$ as $x \to 0$. It is, however, easily proved, by establishing contradictions in the alternative case*, that as $x \to 0$, $\phi(x) \to 0$.

We therefore have to consider solutions of $f(x, y) = 0$, of the type $y = \phi(x)$ or $x = \phi(y)$, for which $\phi \to 0$ as $x \to 0$, or $y \to 0$, and ϕ is continuous near $x = 0$ or $y = 0$. Such a solution we may call *a continuous solution*, or from the geometrical point of view, a **branch** of the curve.

---

* Hardy, *P. M.*, p. 192.

§ 6·310. **The fundamental property of branches through a singular point.** The necessary discussion is simplified by establishing in the first place a certain general property of any branch (*i.e.* continuous solution) of $f(x, y) = 0$, without reference to questions of existence. Questions of existence are treated later on.

THEOREM 6·311. *Any branch must touch at $(0, 0)$ one of the straight lines defined by*

$$\phi_n = 0,$$

where n is the order of the lowest order partial differential coefficient of $f(x, y)$ which does not vanish near $(0, 0)$.

The straight lines defined by $\phi_n = 0$ may all be put in one or other of the forms

$$y = \lambda x,$$

where λ is a root of

$$(6·312) \qquad k_n \lambda^n + j_n \lambda^{n-1} + \ldots + a_n = 0,$$

or

$$x = \lambda' y,$$

where λ' is a root of

$$(6·313) \qquad a_n \lambda'^n + b_n \lambda'^{n-1} + \ldots + k_n = 0.$$

If y is any branch of $f(x, y) = 0$, it is required to show that either

$$(6·314) \qquad y \sim \lambda x,$$

where λ is a root of 6·312, or

$$(6·315) \qquad x \sim \lambda' y,$$

where λ' is a root of 6·313. In the case of $\lambda = 0$ or $\lambda' = 0$ the corresponding relation must be replaced by $y = o(x)$ or $x = o(y)$.

The theorem may be proved by establishing contradictions in every alternative case, for if neither 6·314 nor 6·315 is satisfied, then either

$$(i) \qquad y/x \to \mu \ \ or \ \ x/y \to \mu,$$

where μ is some finite number, zero included, which is not a root of either 6·312 or 6·313,

or (ii) there exist a constant μ, not a root of 6·312 or 6·313, and an infinite sequence of values of x, which tend to 0, such that for these values

$$y/x = \mu.$$

In either case it may be proved that* $\phi_n \sim A x^n$ $(A \neq 0)$ and at the same time* $R_n = O(x^{n+1})$, and since $\phi_n + R_n = 0$ these relations are contradictory.

---

* It may of course be necessary to replace x by y in both these relations.

§ 6·40. **Singular points of the second order. Existence of branches.** We now proceed to consider questions of existence, starting with the simplest case of a singular point of the second order. This case is governed by the homogeneous quadratic form

$$\phi_2 \equiv a_2 x^2 + b_2 xy + c_2 y^2.$$

Let

(6·41)$$\qquad\qquad\qquad \Delta = b_2{}^2 - 4 a_2 c_2.$$

If $\Delta < 0$, the straight lines represented by $\phi_2 = 0$, which provide all the possible tangents to the curve at the origin, are imaginary; if $\Delta > 0$, they are real and distinct; if $\Delta = 0$, ϕ_2 is a perfect square and $\phi_2 = 0$ represents only one straight line.

The nature of the singularity is described by the following theorem.

THEOREM 6·42. **Second order singularities.**

(1) *If* $\Delta < 0$, *the singularity is* **an isolated point.** *There is no branch (real) of the curve* $f(x, y) = 0$ *through the singularity.*

(2) *If* $\Delta > 0$, *the singularity is* **a double point with distinct tangents***. *The curve has two branches through the singularity, one touching each of the distinct lines defined by* $\phi_2 = 0$.

(3) *If* $\Delta = 0$, *the form of the curve is still uncertain but, in general, unless a particular condition is satisfied, the singularity is* **a cusp of the first species.** *The curve has two branches which tend to the singularity from one side only, and have as their common tangent there the line defined by* $\phi_2 = 0$. *The branches lie on opposite sides of the tangent.*

(1) $\Delta < 0$. This case is covered by Theorem 6·311; for since there is no real line which the branch can touch there can be no branch through the singularity.

(2) $\Delta > 0$. In this case there are two distinct real roots, and we may suppose (§ 6·20) that

$$\phi_2 \equiv xy\,;$$

the necessary axes may be oblique. Any branch must either touch the line $y = 0$, satisfying $y = o(x)$, or touch $x = 0$, satisfying $x = o(y)$. Consider the former case and put $y = ux$, so that $u \to 0$. The equation $f(x, y) = 0$ transforms into

$$x^2 u + x^3 F_1(x, u) = 0$$

or

(6·43)$$\qquad\qquad H(x, u) \equiv u + x F_1(x, u) = 0,$$

---

* Frequently called **a node.**

where the function F_1 possesses continuous partial differential coefficients near $(0, 0)$. Moreover

$$H(0, 0) = 0, \quad \left(\frac{\partial H}{\partial u}\right)_0 = 1,$$

and therefore the Existence Theorem $1\cdot51$ may be applied. There exists therefore a unique branch

$$u = \phi_1(x)$$

such that $\phi_1(x) \to 0$ as $x \to 0$. The equation $f(x, y) = 0$ has exactly one branch

$$y = x\,\phi_1(x)$$

touching the line $y = 0$ at the origin. Similarly there is exactly one branch

$$x = y\,\phi_2(y)$$

touching the line $x = 0$ at the origin. There cannot be other branches.

(3) $\Delta = 0$. In this case ϕ_2 is a perfect square and we may suppose (§ $6\cdot20$) that

$$\phi_2 = y^2,$$
$$f(x, y) \equiv y^2 + (a_3 x^3 + b_3 x^2 y + c_3 x y^2 + d_3 y^3) + R_3.$$

Any branch must touch $y = 0$ at the origin. We therefore write $y = ux$, where $u \to 0$, divide by x^2, and obtain

$$(6\cdot44) \quad H(x, u) \equiv u^2 + x(a_3 + b_3 u + c_3 u^2 + d_3 u^3) + x^2 F_1(x, u) = 0,$$

where $F_1(x, u)$ possesses partial differential coefficients near $(0, 0)$.

If $a_3 \neq 0$, the Existence Theorem $1\cdot51$ applies, and establishes the existence of a unique function ψ such that, near $(0, 0)$,

$$x = u^2 \psi(u) \quad (\psi(0) = a \neq 0).$$

Making the substitution $x/a = t^2$, it is easily seen that the origin is an ordinary double point with distinct tangents for the (t, u) curve. There exist, therefore, two branches of the (u, x) curve of the form

$$u = (x/a)^{\frac{1}{2}} g\{(x/a)^{\frac{1}{2}}\}, \quad u = -(x/a)^{\frac{1}{2}} g\{-(x/a)^{\frac{1}{2}}\};$$

and so two branches of the (y, x) curve

$$(6\cdot45) \qquad y = \pm\, x\,(x/a)^{\frac{1}{2}} g\{\pm (x/a)^{\frac{1}{2}}\},$$

where $g(0) = 1$. It will be observed that y is real for real x when and only when $x/a > 0$, i.e. on one side only of the origin. Further, y has opposite signs on the two branches near $(0, 0)$, and therefore the two branches lie on opposite sides of the common tangent. Such a point is called a *cusp of the first species*.

§ 6·460. **Discussion of the doubtful case.** When $a_3 = 0$, *i.e.*
$\partial^3 f/\partial x^3 = 0$ for the particular axes chosen, the above discussion does not
apply, and the nature of the singularity remains doubtful. If we return
to equation 6·44 giving $H(x, u)$, regard it as an equation between x
and u and rearrange the terms in a new Taylor's series we obtain

$$(6·461) \quad (u^2 + b_3 ux + a_4 x^2) + (a_5 x^3 + b_4 x^2 u + c_3 xu^2) + R_3(x, u) = 0.$$

We are concerned with branches through $(0, 0)$ and 6·461 shows
that $(0, 0)$ is a singular point of the second order on the (u, x) curve.
We therefore apply *de novo* the preceding discussion.

Case (1). If the quadratic form $u^2 + b_3 ux + a_4 x^2$ has no real roots,
the origin is an *isolated point* on the (u, x) curve and therefore also on
the (y, x) curve.

Case (2). If this quadratic form has two distinct real roots a and β,
there are two distinct branches of the (u, x) curve passing through $(0, 0)$
and such that

$$u \sim ax, \quad u \sim \beta x$$

respectively. There are therefore two distinct branches of the (y, x)
curve through $(0, 0)$ such that

$$y \sim ax^2, \quad y \sim \beta x^2.$$

It is easy to see that if one root (a say) is zero, the corresponding
branch of the (y, x) curve takes the form

$$y \sim a' x^r \quad (r \geqslant 3).$$

It will be observed that both of these branches are real on both sides
of the origin, which may be called in this case **a double point with
coincident tangents.** The two branches are distinguished by their
necessarily different curvature at the origin.

Case (3). If the quadratic form is a perfect square $(u - ax)^2$, we
again return to the doubtful case. If, however, as is in general the case,
$a_5 \neq 0$ in 6·461, the (u, x) curve, after the last section, has two branches
near $(0, 0)$ touching $u - ax = 0$ and such that

$$u - ax \sim \pm x(\beta x)^{\frac{1}{2}} \quad (\beta \neq 0).$$

It follows that the (y, x) curve has two branches near $(0, 0)$ such that

$$y - ax^2 \sim \pm x^2 (\beta x)^{\frac{1}{2}}.$$

These two branches exist only on one side of $(0, 0)$, have (of course)
the common tangent $y = 0$, but now lie *on the same side of the tangent.*
Such a point is called **a cusp of the second species.** Such a cusp is

an essentially more complicated singularity than a cusp of the first species.

In the exceptional case, however, $a_5 = 0$, and the nature of the curve is still doubtful, for the origin is still a double point of the curve obtained by transformation from the (u, x) curve. It is necessary to start the discussion yet again at the beginning, and to continue until a decision is reached. If the curve is algebraic $f(x, y)$ is a polynomial, and the process must eventually terminate, for each step consumes more terms. If, however, $f(x, y)$ is not a polynomial, the process may never terminate, and the nature of the singularity remains undecided. In all cases in which the process terminates (whatever the stage) the resulting singularity may be classified as one or other of an isolated point, a double point (with distinct or coincident tangents), or a cusp of the first or second species.

§ 6·60. **Singular points of order** n. The discussion of such points is very similar to the case $n = 2$, and may be rapidly sketched. We have already proved that any branch touches one of the lines defined by $\phi_n = 0$. We have therefore only to consider the existence and form of the branch or branches associated with any given factor of ϕ_n. Let the given factor be y, of multiplicity $k \leqslant n$. Then $f(x, y) = 0$ takes the form

$$y^k \psi_{n-k}(x, y) + \phi_{n+1}(x, y) + R_{n+1} = 0.$$

Put $y = ux$, so that $u \to 0$ for the branches in question, and

$$u^k \psi_{n-k}(1, u) + x \phi_{n+1}(1, u) + R'_{n+1} = 0,$$

where $$\psi_{n-1}(1, 0) \neq 0, \quad R'_{n+1} = O(x^2).$$

Case (1). $k = 1$. The origin is an ordinary point of the (u, x) curve and Theorem 1·51 applies. *There exists exactly one branch touching each line defined by a simple factor of* ϕ_n.

Case (2). $k > 1$, *but* $\phi_{n+1}(1, 0) \neq 0$. Theorem 1·51 applies and shows that there exists a unique function of u such that

$$x = u^k \phi(u) \quad (\phi(0) \neq 0).$$

If k is odd arguments similar to those used in Theorem 6·42, Case (3), show that the relation between x and u can be put in the form

$$u = x^{1/k} g(x^{1/k}) \quad (g(0) \neq 0),$$

so that u is real for real x near $(0, 0)$ and changes sign with x. Hence

$$y = x^{1+1/k} g(x^{1/k}),$$

and y does not change sign with x. There is thus one branch corresponding to this factor of ϕ_n which touches it at the origin *without an inflexion*.

If k is even, we obtain in the same way

$$u = \pm (ax)^{1/k} g \{\pm (ax)^{1/k}\} \quad (a = 1/\phi(0)).$$

Here u is real only when x has the sign of a, and y has opposite signs on the two branches. There are therefore two branches to the (y, x) curve corresponding to this factor of ϕ_n, which touch it and form a *cusp of the first species at the origin*.

Case (3). $k > 1$ *and* $\phi_{n+1}(1, 0) = 0$. In this case the (u, x) curve has a singularity of order not greater than k at the origin $(k \leqslant n)$. We make a fresh start to analyse the (u, x) singularity, and proceed as before. If the curve is algebraic, the process will terminate at some stage, since $k \leqslant n$ and each step consumes more terms.

EXAMPLES VI

(1) The various species of singular points of the second order are illustrated by the following algebraic curves at the origin of coordinates.

Isolated point: $x^2 + y^2 + x^3 = 0$.

Double point with distinct tangents: $x^3 + y^3 - 3axy = 0$.

Cusp of first species: $y^2 - x^3 = 0$.

Double point with coincident tangents: $y^2(1 + x) - x^4 = 0$.

Cusp of the second species: $(y - x^2)^2 - x^5 = 0$.

(2) *Singularities of* $x = \phi_1(t)$, $y = \phi_2(t)$. A singular point is one at which $\phi_1'(t) = \phi_2'(t) = 0$. It may be assumed that the point corresponds to $t = 0$, and that $\phi_1(0) = \phi_2(0) = 0$. By suitable change of axes we can arrange that

$$x = at^p f_1(t) \quad (p \geqslant 2, f_1(0) = 1),$$
$$y = bt^{p+a} f_2(t) \quad (a > 0, f_2(0) = 1).$$

Any existent branch touches $y = 0$.

Case (1). p *odd*. The relation between x and t can be replaced by a unique relation

$$t = \left(\frac{x}{a}\right)^{1/p} g \left\{\left(\frac{x}{a}\right)^{1/p}\right\} \quad (g(0) = 1),$$

so that t changes sign with x. We have also

$$y = b \left(\frac{x}{a}\right)^{1+a/p} h \left\{\left(\frac{x}{a}\right)^{1/p}\right\} \quad (h(0) = 1).$$

If a is odd, y does not change sign with x. If a is even, y changes sign with x. In either case, there is one branch of the curve through $(0, 0)$ touching $y = 0$, with in the latter case $y = 0$ for an inflexional tangent. There is apparently no singularity, but really one of a concealed nature, for $d^2y/dx^2 \to \infty$ as $x \to 0$, if $a < p$, and whatever value a has, some differential coefficient is discontinuous at $(0, 0)$.

Case (2). *p even.* We have

$$t = \pm \left(\frac{x}{a}\right)^{1/p} g \left\{ \pm \left(\frac{x}{a}\right)^{1/p} \right\},$$

$$y = (\pm)^{\alpha + p} b \left(\frac{x}{a}\right)^{1 + \alpha/p} h \left\{ \pm \left(\frac{x}{a}\right)^{1/p} \right\}.$$

Thus t and y only exist for values of x on one side of $(0, 0)$, such that $x/a > 0$. There are two branches touching $y = 0$ at $(0, 0)$ and forming a cusp of the first species if α is odd and a cusp of the second species if α is even, provided it is not the case that $h(\lambda)$ is an even function of λ. In this latter case, there is only one branch when α is even.

(3) *Radius of curvature at a cusp.*

Let P be a cusp, Q a point which tends to P along either branch of the curve, and ρ the radius of curvature at Q. Then as $Q \to P$, $\rho \to 0$, if P is the simplest type of cusp of first species.

On the other hand ρ *usually* has a finite limit (different from zero) if P is the simplest type of cusp of the second species though exceptionally $\rho \to \infty$. Consider the more complicated cases.

(4) Discuss the form of the evolute of a curve near the point corresponding to a stationary value of ρ $(d\rho/ds = 0)$ on the original curve. Show that in general the evolute has a cusp of the first species.

(5) Show that in general a cusp of the second species on a given curve corresponds to a point of inflexion on the evolute.

CHAPTER VII

ASYMPTOTES OF PLANE CURVES*

§ 7·10. Definition of "$P \to \infty$." *If $P(x, y)$ is a point on the curve $f(x, y) = 0$, and if P moves along the curve so that one at least of x and y tends to $+ \infty$ or to $- \infty$, then P is said to tend to infinity, and we write*

$$P \to \infty.$$

Definition of an asymptote. *If P be a point on the curve $y = f(x)$ and $P \to \infty$, and if the shortest distance of P from the curve† (or a branch of it)*

$$g(x, y) = 0$$

tends to 0 as $P \to \infty$, then the curve (or the branch of the curve) $g(x, y) = 0$ is said to be an **asymptote** *of the curve $y = f(x)$.*

---

* d.l.V.P., Vol. ii, pp. 391–393.

† This shortest distance will exist provided the curve (or branch) is continuous.

If the curve $g(x, y) = 0$ is a straight line

$$ax + by + c = 0,$$

then this straight line is said to be *a rectilinear asymptote of* $y = f(x)$; this name may be shortened to *asymptote* when there is no possibility of confusion.

§ 7·20. **Properties of asymptotes.** The following theorems explain the definition and are of general utility in obtaining asymptotes to a given curve.

THEOREM 7·21. *In order that* $g(x, y) = 0$ *may be an asymptote to the curve* $y = f(x)$ *it is sufficient that, as* $P \to \infty$ *along* $y = f(x)$, *the distance of* P *to* $g(x, y) = 0$, *measured parallel to a fixed direction, should tend to zero.*

For such a distance is certainly not less than the shortest distance, which therefore tends to zero.

In the particular case of a rectilinear asymptote, which is the important case, this condition is also *necessary**, if the proviso be inserted that the fixed direction is not parallel to the given straight line $g(x, y) = 0$; for then this oblique distance bears a constant finite ratio to the shortest distance.

THEOREM 7·22. *In order that the straight line* $x = a$ *may be an asymptote to the curve* $y = f(x)$ *(with coordinates rectangular or oblique), it is necessary and sufficient that*

$$|f(x)| \to \infty,$$

either as $x \to a + 0$, *or as* $x \to a - 0$†.

The two cases do not need separate treatment. Suppose $x \to a - 0$. If P be the point (x, y) on the curve, then $P \to \infty$ as $x \to a - 0$. Moreover, the shortest distance from P to $x - a = 0$ is $(a - x) \sin \omega$, where ω is the angle between the coordinate axes, and therefore tends to 0 as $P \to \infty$. Thus the condition is sufficient, and it is plainly also necessary.

A similar theorem could be given for asymptotes of the type $y = b$, but this case is covered by the next theorem.

THEOREM 7·23. *In order that the curve* $y = g(x)$ *should be an asymptote of the curve* $y = f(x)$, *it is sufficient (but not necessary) that*

$$f(x) - g(x) \to 0$$

as $x \to \infty$.

---

* The condition is not necessary for general asymptotes. See Theorem 7·23.

† Or both, of course. $x \to a + 0$ means that $x > a$, and $x \to a$, while $x \to a - 0$ means that $x < a$, and $x \to a$.

This is merely a simplification of Theorem 7·21, so far as the sufficiency of the condition is concerned. To show that the condition is not necessary, we have only to prove that, *e.g.*, if $a > 1$, $y = x^a$ is asymptotic to $y = x^a + 1$. We have, in the first quadrant, for the former curve $x = + y^{1/a}$, and for the latter $x = (y + 1)^{1/a}$. Hence the difference between the curves measured parallel to the x-axis, for a given value of y, is

$$(y + 1)^{1/a} - y^{1/a},$$

which tends to 0 as $y \to \infty$, if $1/a < 1$.

In the case of rectilinear asymptotes, however, when $g(x) \equiv cx + d$, this condition is necessary as well as sufficient, as we have already stated.

Before passing on, it is well to state explicitly the following almost obvious facts, which are constantly used in the following sections.

If the straight line $y = cx + d$ is an asymptote to the curve $y = f(x)$ then

$$f(x) \sim cx, \quad f(x) - cx \to d$$

and conversely.

§ 7·240. **Asymptotes as the limits of tangents or chords.** An asymptote is sometimes defined as the limit of a tangent whose point of contact tends to infinity. Asymptotes however may exist in the sense of the definition of § 7·10 when the tangent has no limit*, as its point of contact tends to infinity. This fact shows that "the limit of the tangent" is not a suitable definition of an asymptote. The precise relations between asymptotes and tangents are defined by the following theorems.

THEOREM 7·241. *If a tangent to the curve*

$$y = f(x),$$

whose point of contact is P, has the limit

$$y = cx + d$$

as $P \to \infty$, then $y = cx + d$ is an asymptote.

The tangent at the point $P(\xi, \eta)$ is

$$y - f'(\xi) x - \{\eta - \xi f'(\xi)\} = 0.$$

Suppose for example that, as $\xi \to + \infty$, this tangent tends to the limiting position

$$y - cx - d = 0 ;$$

i.e. suppose that

$$f'(\xi) \to c, \quad f(\xi) - \xi f'(\xi) \to d.$$

---

* The curve need not even have a tangent.

If $c \neq 0, f'(\xi) \sim c$, and therefore* $f(\xi) \sim c\xi$. Therefore

$$\frac{1}{\xi} - \frac{f'(\xi)}{f(\xi)} - \frac{d}{\xi f(\xi)} = o\left\{\frac{1}{\xi f(\xi)}\right\},$$

$$\frac{1}{\xi} - \frac{f'(\xi)}{f(\xi)} - \frac{d}{c\xi^2} = o\left(\frac{1}{\xi^2}\right).$$

On integrating we have

$$A + \log \xi - \log f(\xi) + \frac{d}{c\xi} = o\left(\frac{1}{\xi}\right),$$

or

$$\log\left[\frac{B\xi \exp(d/c\xi)}{f(\xi)}\right] = o\left(\frac{1}{\xi}\right),$$

$$\frac{B\xi \exp(d/c\xi)}{f(\xi)} = 1 + o\left(\frac{1}{\xi}\right).$$

But, as $f(\xi) \sim c\xi$, we must have $B = c$. Therefore

$$\frac{c\xi}{f(\xi)}\left\{1 + \frac{d}{c\xi} + O\left(\frac{1}{\xi^2}\right)\right\} = 1 + o\left(\frac{1}{\xi}\right),$$

and so finally

$$f(\xi) = c\xi + d + o(1);$$

or in other words

$$y = cx + d$$

is an asymptote. The case $c = 0$ alone remains to be considered.

In this case we have to prove that if

$$f'(\xi) \to 0, \quad f(\xi) - \xi f'(\xi) \to d,$$

then $f(\xi) \to d$. Let us write

$$f(\xi) = g(\xi) - \xi;$$

then

$$g'(\xi) \to 1, \quad g(\xi) - \xi g'(\xi) \to d.$$

Therefore, by what we have already proved,

$$g(\xi) = \xi + d + o(1),$$

so that

$$f(\xi) = d + o(1),$$

as was to be proved. It follows that $y = d$ is an asymptote and the proof of our theorem is completed.

If the tangent has a limiting position, it follows from the proof of the last theorem that

$$(7\cdot2411) \qquad\qquad f'(\xi) = c + o\left(\frac{1}{\xi}\right).$$

This condition is however not in itself sufficient to imply the existence

---

* By l'Hospital's theorem, d.l.V.P., Vol. I, p. 124.

of a limit. For example, if $f(x) = cx + \log \log x$, $f'(\xi)$ satisfies 7·2411, but no asymptote $y = cx + d$ exists.

The geometrical meaning of this necessary condition 7·2411 may be mentioned in passing. It is that, unless the slope of the tangent tends sufficiently rapidly to its limit, the point of intersection of the tangent and the y-axis for example will not remain within a finite distance of the origin and "the limit of the tangent will lie wholly at infinity".

THEOREM 7·242. *Under the same conditions as in the last theorem, the asymptote is the limiting position of a chord two of whose points of intersection with the curve, say P, Q, together tend to infinity in a direction in which the curve is asymptotic to its asymptote.*

Let P be (ξ_1, η_1) and Q be (ξ_2, η_2), and suppose that

$$\xi_1 \to +\infty, \quad \xi_2 \to +\infty.$$

Then the chord is

$$y - \eta_1 = \frac{\eta_2 - \eta_1}{\xi_2 - \xi_1}(x - \xi_1)$$

$$= f'(\xi)(x - \xi_1),$$

where $\xi_2 > \xi > \xi_1$, by the mean value theorem, supposing that $\xi_2 > \xi_1$. Since $f'(\xi) \to c$, we have only to prove that

$$\xi_1(f'(\xi) - c) \to 0.$$

But this follows at once from the facts that (by 7·2411)

$$\xi(f'(\xi) - c) \to 0$$

and $\xi_1 < \xi$, and so the theorem is proved.

The restrictive hypothesis needed in the two last theorems is of some interest. Its necessity may be illustrated by the curve

$$y = f(x) \equiv cx + x^{-a} \sin(x^2).$$

If $a > 0$, $y = f(x)$ has the asymptote $y = cx$ as $x \to \infty$, but $f'(x)$ has no limit as $x \to \infty$ unless $a > 1$. If $a > 1$,

$$f'(x) - c = 2x^{1-a} \cos(x^2) - ax^{-1-a} \sin(x^2) \to 0,$$

but

$$x(f'(x) - c) \to 0$$

if and only if $a > 2$. In this case only $(a > 2)$ the tangent tends to the asymptote. This example illustrates what we stated above, namely that the "limit of a tangent whose point of contact P is such that $P \to \infty$" or "the limit of a chord etc." are unsuitable as definitions of a rectilinear asymptote except perhaps for algebraic curves, where the restrictions are always satisfied.

§ 7·250. The preceding definition and theorems have been stated as referring to a curve $y = f(x)$. They all apply without change to any branch of the curve $f(x, y) = 0$ which satisfies the conditions of the implicit function theorem, and so can be expressed in the form

$$y = \psi(x) \ \ (x > x_0), \quad or \quad x = \psi(y) \ \ (y > y_0).$$

As some such conditions are essential to enable us to assert the existence of the branch in question, they form no restriction on the generality of the foregoing discussion.

For a curve given in polar coordinates by the equation $r = f(\theta)$, the fundamental theorem is the following.

THEOREM 7·251. *If $r \to \infty$ as $\theta \to a$ and if*

$$f(\theta)(a - \theta) \to b,$$

then the straight line

$$r \sin (a - \theta) = b$$

is an asymptote to the curve; and conversely.

The proof is left to the reader.

§ 7·30. **Asymptotes of algebraic curves.** In the case of an algebraic curve, the behaviour of any branch as $x \to \infty$ or (and) $y \to \infty$ can always be reduced, by a suitable substitution, to the study of the branches of an algebraic curve in the neighbourhood of the origin. For example, suppose both $x \to \infty$ and $y \to \infty$. The equation of the algebraic curve can be written in the form

$$(7·301) \qquad \phi_n(x, y) + \phi_{n-1}(x, y) + \ldots + \phi_0 = 0,$$

where $\phi_r(x, y)$ is a polynomial in x and y, homogeneous and of degree r. Write $x = 1/x'$, $y = 1/y'$. Then 7·301 in general transforms into

$$\frac{1}{(x'y')^n} \phi_n(y', x') + \frac{1}{(x'y')^{n-1}} \phi_{n-1}(y', x') + \ldots + \phi_0 = 0,$$

or

$$\phi_n(y', x') + x'y' \phi_{n-1}(y', x') + \ldots + (x'y')^n \phi_0 = 0.$$

We have therefore to study the form of the branches of this curve in the neighbourhood of the origin, which is a multiple point of order n. This is the problem whose solution was sketched in § 6·60. It is convenient however to develop an alternative direct method of attack, though it should always be borne in mind that the study of the asymptotes of algebraic curves is identical theoretically with the study of the form of the branches through a singular point.

When we discuss the existence of branches belonging to any possible asymptote, we shall in general reduce the problem by a substitution to the problem of the existence of a branch touching a definite straight line through the origin; for to any chosen asymptote a definite tangent at the origin is made to correspond by the substitution. This provides a convenient way of specifying precisely what is meant by a single branch or branches of the algebraic curve near infinity. The branch associated with a particular asymptote means the branch corresponding to that which touches a particular tangent at the origin of the transformed curve.

Suppose that $f(x, y) = 0$ is the equation of an algebraic curve of degree n, which may be put into the form 7·301. We shall attempt to find the rectilinear asymptotes of the various branches of this curve which are not parallel to the axis of y, $i.e.$ asymptotes of the form

$$y = cx + d.$$

It is clear that an exactly similar procedure, on interchanging x and y, will find for us the asymptotes of the form

$$x = cy + d,$$

$i.e.$ those not parallel to the axis of x. The apparent exception does therefore not limit the generality of the discussion.

As $P \to \infty$ along a branch of the curve $f(x, y) = 0$ in a direction ($x \to +\infty$ say) asymptotic to the asymptote $y = cx + d$, we have

$$y \sim cx, \quad y - cx \to d.$$

Writing $y = tx$, $x = 1/x'$ in equation 7·301 and dividing through by x^n, we have

$$(7\text{·}302) \qquad \phi_n(1, t) + x' \phi_{n-1}(1, t) + \ldots = 0,$$

an equation giving t in terms of x'.

Now by hypothesis $t \to c$ as $x \to \infty$, $i.e.$ as $x' \to 0$, so that t remains finite as $x' \to 0$. Moreover, it satisfies an equation of the form

$$a_0 t^\lambda + a_1 t^{\lambda-1} + \ldots + (a_\lambda + o(1)) = 0, \quad (\lambda \leqslant n).$$

Now the roots of such an equation are continuous functions of a_λ the constant term. Therefore, as $x' \to 0$, $t \to a$, where a is the root of the equation

$$a_0 t^\lambda + \ldots + a_\lambda = 0,$$

$i.e.$ of

$$(7\text{·}303) \qquad \phi_n(1, a) = 0.$$

We have therefore found that *the first condition to be satisfied by the asymptote $y = cx + d$ is*

$$(7\text{·}31) \qquad \phi_n(1, c) = 0.$$

Unless this equation has real roots there will be no asymptotes of the assumed form. Let us suppose therefore that c is a real root of this equation. Writing $y = cx + v = v + c/x'$, and substituting for y in the equation of the curve, we have

$$\phi_n(x, cx + v) + \phi_{n-1}(x, cx + v) + \ldots = 0.$$

It is required that $v \to d$, as $x \to \infty$. On dividing through this equation by x^n, we have

$$\phi_n(1, c + x'v) + x' \phi_{n-1}(1, c + x'v) + \ldots = 0,$$

and, by applying Taylor's theorem,

$$x'v \, \phi_n'(1, c) + \tfrac{1}{2} x'^2 v^2 \phi_n''(1, c) + \ldots + x' \phi_{n-1}(1, c) + x'^2 v \phi_{n-1}'(1, c) + \ldots = 0.$$

The number of terms in each expansion is finite, so that there is no question of validity. Suppose now that

$$\phi_n{}'(1, c) \neq 0,$$

i.e. that c is a simple root of the equation $\phi_n(1, t)=0$. Then v satisfies the equation

$$v\phi_n{}'(1, c) + \phi_{n-1}(1, c) + x'\psi(v, x') = 0,$$

where ψ is a polynomial in v and x' of degree n at most. Since $v \to d$, v remains finite as $x' \to 0$, and therefore

$$\psi(v, x') = O(1).$$

Therefore, by a repetition of our preceding arguments, d must be a root of the equation

$$d\phi_n{}'(1, c) + \phi_{n-1}(1, c) = 0,$$

i.e.

$$(7\cdot32) \qquad\qquad d = -\phi_{n-1}(1, c)/\phi_n{}'(1, c)$$

provided $\phi_n{}'(1, c) \neq 0$. We have therefore found that *if $y=cx+d$ is an asymptote then*

$$\phi_n(1, c) = 0,$$

and if c is a simple real root of $\phi_n(1, c)=0$ *then*

$$d = -\phi_{n-1}(1, c)/\phi_n{}'(1, c).$$

We must now consider the case

$$\phi_n{}'(1, c) = 0,$$

in which c is a multiple root of $\phi_n(1, t)=0$. Suppose first of all that

$$\phi_{n-1}(1, c) \neq 0.$$

Then the equation satisfied by v may be written $\phi_{n-1}(1, c) + x'\psi(v, x')=0$, and as v is to remain finite as $x' \to 0$, we have

$$\phi_{n-1}(1, c) + O(x') = 0,$$

which contradicts $\phi_{n-1}(1, c) \neq 0$. *There is therefore in this case no rectilinear asymptote.*

Let us now suppose that

$$\phi_{n-1}(1, c) = 0.$$

Then the equation satisfied by v may be written

$$\frac{1}{2!}x'^2v^2\,\phi_n{}''(1, c) + \frac{1}{3!}x'^3v^3\phi_n{}'''(1, c) + \ldots + x'^2v\phi'_{n-1}(1, c)$$

$$+ \frac{1}{2!}x'^3v^2\phi''_{n-1}(1, c) + \ldots + x'^2\phi_{n-2}(1, c) + \ldots + \ldots = 0,$$

or $\qquad \frac{1}{2}v^2\phi_n{}''(1, c) + v\phi'_{n-1}(1, c) + \phi_{n-2}(1, c) + O(x') = 0.$

It follows that, provided either $\phi_n{}''(1, c) \neq 0$ or $\phi'_{n-1}(1, c) \neq 0$, d must satisfy the quadratic (possibly linear) equation

$$(7\cdot33) \qquad \tfrac{1}{2}d^2\phi_n{}''(1, c) + d\phi'_{n-1}(1, c) + \phi_{n-2}(1, c) = 0\,;$$

that if $\phi_n{}''(1, c)=0$, $\phi'_{n-1}(1, c)=0$, and $\phi_{n-2}(1, c) \neq 0$, there is no rectilinear asymptote of this type; while if $\phi_n{}''(1, c)=\phi'_{n-1}(1, c)=\phi_{n-2}(1, c)=0$, the matter is still uncertain. We then continue the above process and, since the degree of the equation is finite, the process must eventually terminate, leaving us either no asymptotes of the assumed type, or else equations by which to determine c and d. We can sum up the results of the preceding discussion as follows.

In order that

$$y = cx + d$$

may be an asymptote of the curve

$$\phi_n(x, y) + \phi_{n-1}(x, y) + \ldots = 0,$$

it is necessary that c should be a real root of

$$\phi_n(1, t) = 0\,;$$

and that, if c is a simple root of this equation,

$$d = -\phi_{n-1}(1, c)/\phi_n{}'(1, c).$$

If c is a multiple root, there is no asymptote unless

$$\phi_{n-1}(1, c) = 0.$$

If also $\phi_{n-1}(1, c) = 0$, then d must satisfy the equation

$$\tfrac{1}{2}d^2\phi_n{}''(1, c) + d\phi'_{n-1}(1, c) + \phi_{n-2}(1, c) = 0$$

provided this equation contains d; and so on.

A less explicit but more compact statement of these facts may be made as follows. We first of all notice that, on substituting $y = v + c/x'$ into the equation of the curve, and reducing, we eventually obtain in all cases an equation of the form

$$(7\cdot331) \qquad\qquad \psi(v, c) + x'\psi(v, c) + \ldots = 0,$$

where $\psi(v, c)$ is polynomial in v and c which is not identically null, but which may or may not contain v. It follows that d must be determined by the equation

$$(7\cdot332) \qquad\qquad \psi(d, c) = 0,$$

and that if this does not contain d then there is no rectilinear asymptote for this particular value of c. Then we may say that *in order that*

$$y = cx + d$$

may be an asymptote of the curve

$$\phi_n(x, y) + \phi_{n-1}(x, y) + \ldots = 0,$$

it is necessary that c should be a real root of

$$\phi_n(1, t) = 0,$$

and that, when a value of c has thus been determined, d should be a real root of

$$\psi(v, c) = 0.$$

If there is no such real root there are no rectilinear asymptotes for this value of c.

There is yet one more necessary condition that follows from the assertion that $y=cx+d$ is an asymptote, namely that there should exist a real branch of the curve $f(x,y)=0$ of the form

$$y=cx+d+u,$$

where $u \to 0$ as $x \to \infty$ (or as $x \to -\infty$ as the case may be). On substituting $d+u$ for v in equation 7·331 we have

$$(7·333) \qquad \psi(d+u, c)+x'\psi_1(d+u, c)+\ldots=0,$$

or for the sake of shortness

$$\chi(d+u, c, x')=0,$$

and the necessary condition that we require is that equation 7·333 admits at least one real root $u(x')$ such that $u(x') \to 0$ as $x' \to +0$ [or as $x' \to -0$ as the case may be]. The problem is thus reduced to the ordinary problem of the existence of implicit functions, and admits the solutions of Theorem 1·51 or Chapter VI.

It may be noticed that the sign of $u(x')$ determines whether the curve lies above or below its asymptote for large values of x, and that if $u(x')$ exists and $u(x') \to 0$ as $x' \to +0$, or as $x' \to -0$, the curve approaches its asymptote as $x \to +\infty$ or as $x \to -\infty$ respectively.

It is easily seen that, *with the addition of the last condition*, the foregoing necessary conditions for the existence of rectilinear asymptotes are also sufficient. We may therefore enunciate the following theorem, which embodies one of the usual rules for obtaining rectilinear asymptotes.

THEOREM 7·34. **Rule I for Asymptotes.** *In order that*

$$y=cx+d$$

may be an asymptote of the curve

$$\phi_n(x, y)+\phi_{n-1}(x, y)+\ldots=0,$$

it is necessary and sufficient (1) *that c should be a real root of*

$$\phi_n(1, t)=0;$$

(2) *that when c has so been chosen, d should be a real root of*

$$\psi(v, c)=0;$$

and (3) *that when d has so been chosen*

$$\chi(d+u, c, x')=0$$

should admit at least one real root $u(x')$ such that $u(x') \to 0$ as $x' \to +0$ or (and) as $x' \to -0$.

Here $\psi(v, c)$ and $\chi(d+u, c, x')$ are certain polynomials which have been defined in the course of the foregoing discussion. The existence of the branch $u(x')$ may be discussed when necessary by the methods of Chapter VI.

§7·40. **Asymptotes parallel to the axes of coordinates.** It is of interest to consider in greater detail the case in which zero is a possible value of c, *i.e.* in which

$$\phi_n(1, 0)=0.$$

We note that $\phi_n(1, 0)$ is the coefficient of x^n. In this case

$$d = -\phi_{n-1}(1, 0)/\phi_n'(1, 0) \qquad (\phi_n'(1, 0) \neq 0).$$

Now $\phi_n'(1, 0)$ is the coefficient of t in $\phi_n(1, t)$, and $\phi_{n-1}(1, 0)$ is the term independent of t in $\phi_{n-1}(1, t)$. Hence

$$y\phi_n'(1, 0) + \phi_{n-1}(1, 0)$$

is the coefficient of x^{n-1}, i.e. of the highest power of x in $f(x, y)$. Moreover

$$y\phi_n'(1, 0) + \phi_{n-1}(1, 0) = 0$$

is a possible asymptote, and effectively one if condition (3), that a suitable real solution of

$$\chi(d + u, 0, x') = 0$$

exists, is satisfied. Again if $\phi_n'(1, 0) = 0$, and $\phi_{n-1}(1, 0) = 0$, d is determined by the equation (unless meaningless)

$$\tfrac{1}{2}d^2\phi_n''(1, 0) + d\phi'_{n-1}(1, 0) + \phi_{n-2}(1, 0) = 0,$$

and the asymptotes if they exist are given by

$$\tfrac{1}{2}y^2\phi_n''(1, 0) + y\phi'_{n-1}(1, 0) + \phi_{n-2}(1, 0) = 0,$$

where the expression on the left is the coefficient of x^{n-2}, i.e. of the highest power of x in $f(x, y)$. It is easy to see that this holds in general, and therefore that all possible asymptotes of the form $y = d$ may be obtained by equating to zero the coefficient of the highest power of x occurring in the equation $f(x, y) = 0$. It should be observed that this rule can only be effective when the term x^n does not occur. We may therefore state the following theorem embodying this rule.

Theorem 7·41. **Rule II for Asymptotes.** *Rectilinear asymptotes parallel to the axis of y (x) can only exist if the term in y^n (x^n) does not occur in $f(x, y)$. When they exist, they can be obtained by equating to zero the coefficient of the highest power of y (x) occurring in $f(x, y)$. Condition (3) must be shown to be satisfied for each line so obtained, before we may assert that such a line is actually an asymptote.*

§ 7·50. **Existence of branches of the curve, asymptotic to the asymptote.** It is important to call attention to certain general cases in which condition (3) is automatically satisfied, and in which, therefore, when we have obtained c and d we can at once assert that $y = cx + d$ is an asymptote[*]. This we do in the following theorem.

Theorem 7·51. *If d is a simple root of $\psi(v, c) = 0$, or in particular if c is a simple root of $\phi(1, t) = 0$, then*

$$y = cx + d$$

is an asymptote, and moreover asymptotic to a single branch of the curve both as $x \to +\infty$ and as $x \to -\infty$. The branch in general lies on opposite sides of the asymptote at the two ends.

* Cf. § 6·60.

If c is a simple root of $\phi(1, t)=0$, we have already seen that the equation determining d is linear, and therefore that d is a simple root of $\psi(v, c)=0$. When this is so, the equation determining u, namely,

$$\psi(d+u, c)+x'\psi_1(d+u, c)+\ldots=0,$$

may be written

$$\chi(u, x') \equiv u\psi'(d, c)+\tfrac{1}{2}u^2\psi''(d, c)+\ldots+x'\psi_1(d+u, c)+\ldots=0,$$

where $\psi'(d, c)\neq0$, and the values $(0, 0)$ satisfy the equation. The origin of the curve $\chi(u, x')=0$ is therefore an ordinary point, since $\left(\dfrac{\partial\chi}{\partial u}\right)_0\neq0$. It follows by the Existence Theorem 1·51 that a unique function $u(x')$ exists for values of x' such that $|x'|<k$, and that $u(x')\to0$ as $|x'|\to0$. Further $u(x')$ in general changes sign with x'. Condition (3) is satisfied, and our theorem is proved. This corresponds to the case $k=1$ of § 6·60.

We may apply the preceding reasoning to Rule II, and obtain the following theorem analogous to 7·51.

THEOREM 7·511. *If the highest power of x is x^{n-1}, and if its coefficient is $ay+b\ (a\neq0)$, then $ay+b=0$ is an asymptote, asymptotic in both the directions $y\to+\infty$, $y\to-\infty$ to a single branch of the curve.*

We may observe here that the need for condition (3) is simply in order to exclude the possibility of $u(x')$ being complex for real values of x' however small. Since $(0, 0)$ is always a point on the algebraic curve

$$\chi(d+u, c, x')=0,$$

there always exist one or more functions $u(x')$, such that $u(x')\to0$ as $|x'|\to0$, if complex values are admitted. In this case too, complex values of c and d may also be admitted, and to such a pair will correspond a branch of the curve (necessarily complex). The need for condition (3) arises from the fact that while to complex asymptotes correspond complex branches of the curve, to real asymptotes do not necessarily correspond real branches. The corresponding phenomenon in the case of singular points is the occurrence of *an isolated point*.

If we strike out all conditions of reality for real values of x, Theorem 7·34 takes the simpler form.

In order that $y=cx+d$ may be an asymptote of the curve

$$\phi_n(x, y)+\phi_{n-1}(x, y)+\ldots=0$$

it is necessary and sufficient (1) that c should be a root of

$$\phi_n(1, t)=0$$

and (2) that when c has been so chosen d should be a root of

$$\psi(v, c)=0.$$

In general $\phi_n(1, t)=0$ has n distinct roots. We may therefore, with due caution, say that *in general* a curve of degree n has n rectilinear asymptotes real or complex.

§ 7·60. After what precedes, the reader should have no difficulty in giving a strict proof of the validity of the following rule for the rectilinear asymptotes of an algebraic curve.

THEOREM 7·61. **Rule III for Asymptotes.** *If the equation of an algebraic curve can be expressed in the form*

$$f_n(x, y) + f_{n-2}(x, y) = 0,$$

where
$$f_n(x, y) = \prod_{r=1}^{n} (a_r x + b_r y + c_r),$$

and $f_{n-2}(x, y)$ is of degree $n-2$ at most in x and y, and if no factor of $f_n(x, y)$ is a constant, and no two factors of $f_n(x, y)$ represent identical or parallel straight lines, then the curve has n asymptotes whose equations are

$$a_r x + b_r y + c_r = 0 \qquad (r = 1, 2, \dots n).$$

It should be noted that in this case all the lines are necessarily asymptotes to real branches of the curve (Theorem 7·51).

§ 7·70. **Curvilinear asymptotes.** We have seen that under certain conditions, although c is a real root of $\phi_n(1, t) = 0$, yet there cannot exist any corresponding rectilinear asymptote. [In fact in such cases there cannot exist any such asymptote even if complex values are taken into consideration.] An investigation similar to the preceding shows us that in certain cases there exists a real branch of $f(x, y) = 0$, which satisfies the relation $y \sim cx$, as $P \to \infty$, but which does not satisfy the relation $y - cx \to d$ for any finite value of d. In order to obtain a knowledge of the form of any branch of the curve $f(x, y) = 0$ as $P \to \infty$, it is necessary to undertake a further investigation of such cases. We shall find that instead of having a straight line as asymptote, the curve has, in the simplest case, a parabola as asymptote. We proceed to discuss this simplest case before passing on to the general one.

Parabolic asymptotes. The simplest case left over from the last section is that in which

$$\phi_n(1, c) = \phi_n'(1, c) = 0, \quad \phi_{n-1}(1, c) \neq 0;$$

and the simplest form of this case is obtained by supposing

$$\phi_n''(1, c) \neq 0.$$

We write $y = v + cx = v + c/x'$, and note that we must have

$$(7·72) \qquad v = o(x).$$

The curve then takes the form

$$\frac{1}{2!} x'^2 v^2 \phi_n''(1, c) + \frac{1}{3!} x'^3 v^3 \phi_n'''(1, c) + \dots + x' \phi_{n-1}(1, c) + o(x') = 0,$$

or
$$\tfrac{1}{2} v^2 \{\phi_n''(1, c) + o(1)\} + x \{\phi_{n-1}(1, c) + o(1)\} = 0.$$

In order that this equation may have a real solution it is necessary that

$$x \phi_{n-1}(1, c)/\phi_n''(1, c) < 0,$$

i.e. that $x \to -\infty$ if $\phi_{n-1}(1, c)/\phi_n''(1, c) > 0$, and $x \to +\infty$ if

$$\phi_{n-1}(1, c)/\phi_n''(1, c) < 0.$$

It will be sufficient to consider one of these cases, for instance the latter. We put

$$(7·73) \qquad a = \{-2\phi_{n-1}(1, c)/\phi_n''(1, c)\}^{\frac{1}{2}}.$$

Then, as $x \longrightarrow +\infty$, it is necessary that either $v \sim ax^{\frac{1}{2}}$, or $v \sim -ax^{\frac{1}{2}}$, both of which hypotheses fulfil the condition 7·72. We must therefore have

$$y = cx \pm ax^{\frac{1}{2}} (1 + o(1)),$$

as $x \longrightarrow +\infty$, while as $x \longrightarrow -\infty$ there can be no real values of y satisfying 7·72. We have still to go a step further before we find an asymptote to the branch. Putting $v = \pm ax^{\frac{1}{2}} + u$, where $u = o(x^{\frac{1}{2}})$, we find that u must satisfy

$$\pm ax^{\frac{1}{2}} u \{\phi_n''(1,c) + o(1)\} \pm ax^{\frac{1}{2}} [\phi'_{n-1}(1,c) + \tfrac{1}{6}a^2\phi_n'''(1,c) + o(1)] + O(1) = 0,$$

or $\qquad\qquad u \sim -\phi'_{n-1}(1,c)/\phi_n''(1,c) - a^2\phi_n'''(1,c)/6\phi_n''(1,c).$

Therefore *we must have*

$$y = cx \pm ax^{\frac{1}{2}} + \beta + o(1),$$

as $x \longrightarrow \infty$, *where* $\qquad a = \{-2\phi_{n-1}(1,c)/\phi_n''(1,c)\}^{\frac{1}{2}},$

and $\qquad\qquad\qquad \beta = -\phi'_{n-1}/\phi_n'' + \phi_{n-1}\phi_n'''/3\phi_n''^2,$

where ϕ'_{n-1} *etc. stand for* $\phi'_{n-1}(1,c)$, *so that the two branches of the curve (if they exist) must be asymptotic to the two arms of the parabola*

$$(y - cx - \beta)^2 = a^2 x.$$

It only remains to show that if we write

$$y = cx \pm ax^{\frac{1}{2}} + \beta + w,$$

there exists in both cases a real function w for large positive values of x, such that $w \longrightarrow 0$ as $x \longrightarrow +\infty$; this follows at once from the fundamental theorem 1·51 in the manner of 7·50, for β is the root of a linear equation. The reader will have no difficulty in supplying the details. We have therefore proved the following theorem.

THEOREM 7·74. **Parabolic asymptotes.** *If c is a root of $\phi_n(1,c) = 0$ such that*

$$\phi_n(1,c) = \phi_n'(1,c) = 0, \quad \phi_{n-1}(1,c) \neq 0, \quad \phi_n''(1,c) \neq 0,$$

then there exist two branches of the curve

$$\phi_n(x,y) + \phi_{n-1}(x,y) + \ldots = 0$$

which possess the two arms of the parabola

$$(y - cx + \phi'_{n-1}/\phi_n'' - \phi_{n-1}\phi_n'''/3\phi_n''^2)^2 = -2x\,\phi_{n-1}/\phi_n''$$

for parabolic asymptotes.

The best possible parabola. Before leaving this simplest case it should be noticed that the arms of any parabola of the form

$$(y - cx - \beta)^2 = a^2 x + \mu,$$

where μ is any constant, are asymptotic to the two branches of the given curve. We can in fact determine μ in such a way as to give the closest possible approximation to the two branches, and as the parabola is in no way rendered more complicated by a value of μ other than zero, it is worth while to determine this best possible parabola. We shall find that, while in general the shortest distance between the parabola and the curve is $O(x^{-\frac{1}{2}})$, for one and only one value of μ it is $O(1/x)$.

Proceeding as in the last section we find that $w \sim \pm \gamma x^{-\frac{1}{2}}$, and that if $w = \pm \gamma x^{-\frac{1}{2}} + z$ then $z = O\,(1/x)$.

It follows that
$$y - cx - \beta = \pm ax^{\frac{1}{2}} \pm \gamma x^{-\frac{1}{2}} + O\,(1/x).$$

Moreover, if $(y - cx - \beta)^2 = a^2 x + \mu$,
$$y - cx - \beta = \pm ax^{\frac{1}{2}} \pm \tfrac{1}{2}\,(\mu/a)\,x^{-\frac{1}{2}} + O\,(1/x).$$

If therefore
$$\mu = 2a\gamma,$$
the ordinates of the curve and the parabola differ by $O\,(1/x)$, while for all other values of μ the ordinates differ by $O\,(x^{-\frac{1}{2}})$. It follows that *the best possible representation of these branches of the curve by a parabola is afforded by the parabola*
$$(y - cx - \beta)^2 = a^2 x + \mu,$$
where
$$\mu = 2a\gamma.$$

The general case. The following theorem covers all cases.

THEOREM 7·75. **General curvilinear asymptotes.** *If c is a root of*
$$\phi_n\,(1,\,t) = 0$$
to which no rectilinear asymptote can correspond, but to which correspond one or more real branches of the curve
$$\phi_n\,(x,\,y) + \phi_{n-1}\,(x,\,y) + \ldots = 0$$
satisfying
$$y \sim cx,$$
as $x \to +\infty$ or $x \to -\infty$ as the case may be, then any one of these branches takes the form
$$y = cx + dt^p \left\{1 + \sum_{i=1}^{\nu} a_i t^{-i}\right\} + O\,(t^{-1}),$$
where p is a positive integer, $d \neq 0$, and t is determined in one of the following ways :

(1) $$t = x^{1/q} \qquad (q > p),$$
where q is odd, and the curve has one real branch of the given form both as $x \to +\infty$ and as $x \to -\infty$;

(2) $$t = x^{1/q} \qquad (q > p),$$
where q is even, and the curve can have two real branches of the given form as $x \to +\infty$ and none as $x \to -\infty$;

(3) $$t = (-x)^{1/q} \qquad (q > p),$$
where q is even and the curve can have two real branches of the given form as $x \to -\infty$ and none as $x \to +\infty$.

The corresponding asymptote is of course
$$y = cx + dt^p \left\{1 + \sum_{i=1}^{\nu} a_i t^{-i}\right\}.$$

This theorem is obviously merely a natural extension of the fundamental existence theorem, on the lines sketched in Chapter VI, especially §§ 6·40, 6·60. The details may be left to the reader.

NOTE A

A PROPERTY OF DIFFERENTIAL COEFFICIENTS

In pursuance of our general policy we define, for instance, the tangent at $P(x_0)$, to the curve $y = f(x)$, as the limit of the chord PQ when $Q \to P$. For this purpose we require merely the existence of $f'(x_0)$. The important question then arises "when does the chord $Q_1 Q_2$ tend to the tangent at P as $Q_1 \to P$ and $Q_2 \to P$?" Similar questions occur in connection with curvature. [Theorems 2·21, 3·23.]

In general, of course, a more stringent condition than the mere existence of $f'(x_0)$ is required. But cases of *geometrical* interest occur in which actually no more stringent condition is required for two moving points than for one. In the example quoted, if $Q_1 \to P$ and $Q_2 \to P$ from opposite sides, then the chord $Q_1 Q_2$ does tend to the tangent at P provided only $f'(x_0)$ exists.

The cases that occur all reduce to the following question: "*When can it be asserted that the existence of $f'(x_0)$ implies that*

$$\frac{f'(x_1) - f'(x_2)}{x_1 - x_2} \to f'(x_0),$$

as $x_1, x_2 \to x_0$?" The question is answered in a simple manner by the following theorem which it is convenient to state and prove here.

If $f'(0)$ exists, and $|x_1 - x_2|$ is never small compared to the smaller of x_1 and x_2, then*

$$\frac{f(x_1) - f(x_2)}{x_1 - x_2} \to f'(0), \quad \dots\dots\dots\dots\dots\dots\dots(A)$$

as $x_1, x_2 \to 0$. *In particular if x_1 and x_2 have always opposite signs then* (A) *is certainly true.*

Suppose for simplicity that $f(0) = 0$. Then as $x_1 \to 0$, by definition,

$$\frac{f(x_1)}{x_1} = \frac{f(x_1) - f(0)}{x_1 - 0} \to f'(0),$$

since $f'(0)$ exists. Therefore $f(x_1) = x_1 f'(0) + o(x_1)$. Similarly,

$$f(x_2) = x_2 f'(0) + o(x_2),$$

and therefore

$$\frac{f(x_1) - f(x_2)}{x_1 - x_2} = f'(0) + o\{x/|x_1 - x_2|\},$$

where x is the larger of $|x_1|$ and $|x_2|$. If now, as $x_1, x_2 \to 0$, $|x_1 - x_2| > kx$ for some fixed positive value of k, then $o\{x/|x_1 - x_2|\} = o(1/k) \to 0$, and therefore

$$\frac{f(x_1) - f(x_2)}{x_1 - x_2} \to f'(0).$$

---

* *I.e.* there exists a constant k, independent of x_1 and x_2, such that

$$|x_1 - x_2| > k|x_1|, \quad |x_1 - x_2| > k|x_2|.$$

If in particular x_1 and x_2 have opposite signs, which is the geometrically interesting case, then $|x_1 - x_2| > |x_1|$, $|x_1 - x_2| > |x_2|$; the extra condition is automatically satisfied, and the mere existence of $f'(0)$ is sufficient to ensure that

$$\frac{f(x_1) - f(x_2)}{x_1 - x_2} \rightarrow f'(0).$$

NOTE B

THE REMAINDER IN TAYLOR'S THEOREM

We have made use in Note A of the equation $f(x) = xf'(0) + o(x)$, or more generally

$$f(x) = f(0) + xf'(0) + o(x),$$

as equivalent to the existence of $f'(0)$. Similar equations are used frequently in this tract when we wish to avoid unnecessary assumptions, for such equations just contain all the information provided by the hypotheses. In general we can obtain an O- or o-result for the remainder in Taylor's expansion with less assumptions than are required for the use of any of the standard forms of remainder.

For example, suppose that $f^{(n)}(0)$ exists. This is equivalent to the equation

$$f^{(n-1)}(x) = f^{(n-1)}(0) + xf^{(n)}(0) + o(x).$$

Integrating this equation from 0 to x, we obtain, by L'Hospital's theorem,

$$f^{(n-2)}(x) = f^{(n-2)}(0) + xf^{(n-1)}(0) + \tfrac{1}{2}x^2 f^{(n)}(0) + o(x^2).$$

If we repeat the integration $n-2$ times more, we get

$$f(x) = \sum_0^n \frac{1}{r!} x^r f^{(r)}(0) + o(x^n) \; ;$$

in other words we have proved that

$$R_{n+1} = o(x^n).$$

For this o-result, $R_{n+1} = o(x^n)$, the only hypothesis required is the existence of $f^{(n)}(0)$. In the same way we can prove that *for the O-result, $R_{n+1} = O(x^{n+1})$, the only hypothesis required is that all the upper and lower derivates of $f^{(n)}(x)$ should be bounded at $x = 0$.*

To obtain a result at least as good as either the o-result or the O-result from, for example, Lagrange's remainder form, we must assume that $f^{(n+1)}(x)$ exists over an interval containing $x = 0$.

A series of hardcover reprints of major works in mathematics, science and engineering.
All editions are 5⅝ × 8½ unless otherwise noted.

Mathematics

Theory of Approximation, N. I Achieser. Unabridged republication of the 1956 edition. 320pp. 49543-4

The Origins of the Infinitesimal Calculus, Margaret E. Baron. Unabridged republication of the 1969 edition. 320pp. 49544-2

A Treatise on the Calculus of Finite Differences, George Boole. Unabridged republication of the 2nd and last revised edition. 352pp. 49523-X

Space and Time, Emile Borel. Unabridged republication of the 1926 edition. 15 figures. 256pp. 49545-0

An Elementary Treatise on Fourier's Series, William Elwood Byerly. Unabridged republication of the 1893 edition. 304pp. 49546-9

Substance and Function & Einstein's Theory of Relativity, Ernst Cassirer. Unabridged republication of the 1923 double volume. 480pp. 49547-7

A History of Geometrical Methods, Julian Lowell Coolidge. Unabridged republication of the 1940 first edition. 13 figures. 480pp. 49524-8

Linear Groups with an Exposition of Galois Field Theory, Leonard Eugene Dickson. Unabridged republication of the 1901 edition. 336pp. 49548-5

Continuous Groups of Transformations, Luther Pfahler Eisenhart. Unabridged republication of the 1933 first edition. 320pp. 49525-6

Transcendental and Algebraic Numbers, A. O. Gelfond. Unabridged republication of the 1960 edition. 208pp. 49526-4

Lectures on Cauchy's Problem in Linear Partial Differential Equations, Jacques Hadamard. Unabridged reprint of the 1923 edition. 320pp. 49549-3

The Theory of Branching Processes, Theodore E. Harris. Unabridged, corrected republication of the 1963 edition. xiv+230pp. 49508-6

The Continuum, Edward V. Huntington. Unabridged republication of the 1917 edition. 4 figures. 96pp. 49550-7

Lectures on Ordinary Differential Equations, Witold Hurewicz. Unabridged republication of the 1958 edition. xvii+122pp. 49510-8

Mathematical Methods and Theory in Games, Programming, and Economics: Two Volumes Bound as One, Samuel Karlin. Unabridged republication of the 1959 edition. 848pp. 49527-2

Famous Problems of Elementary Geometry, Felix Klein. Unabridged reprint of the 1930 second edition, revised and enlarged. 112pp. 49551-5

Lectures on the Icosahedron, Felix Klein. Unabridged republication of the 2nd revised edition, 1913. 304pp. 49528-0

On Riemann's Theory of Algebraic Functions, Felix Klein. Unabridged republication of the 1893 edition. 43 figures. 96pp. 49552-3

A Treatise on the Theory of Determinants, Thomas Muir. Unabridged republication of the revised 1933 edition. 784pp. 49553-1

A Survey of Minimal Surfaces, Robert Osserman. Corrected and enlarged republication of the work first published in 1969. 224pp. 49514-0

The Variational Theory of Geodesics, M. M. Postnikov. Unabridged republication of the 1967 edition. 208pp. 49529-9